D0384604

IN THE COUNTRY OF *Gazelles*

IN THE COUNTRY OF *Gazelles*

FRITZ R. WALTHER

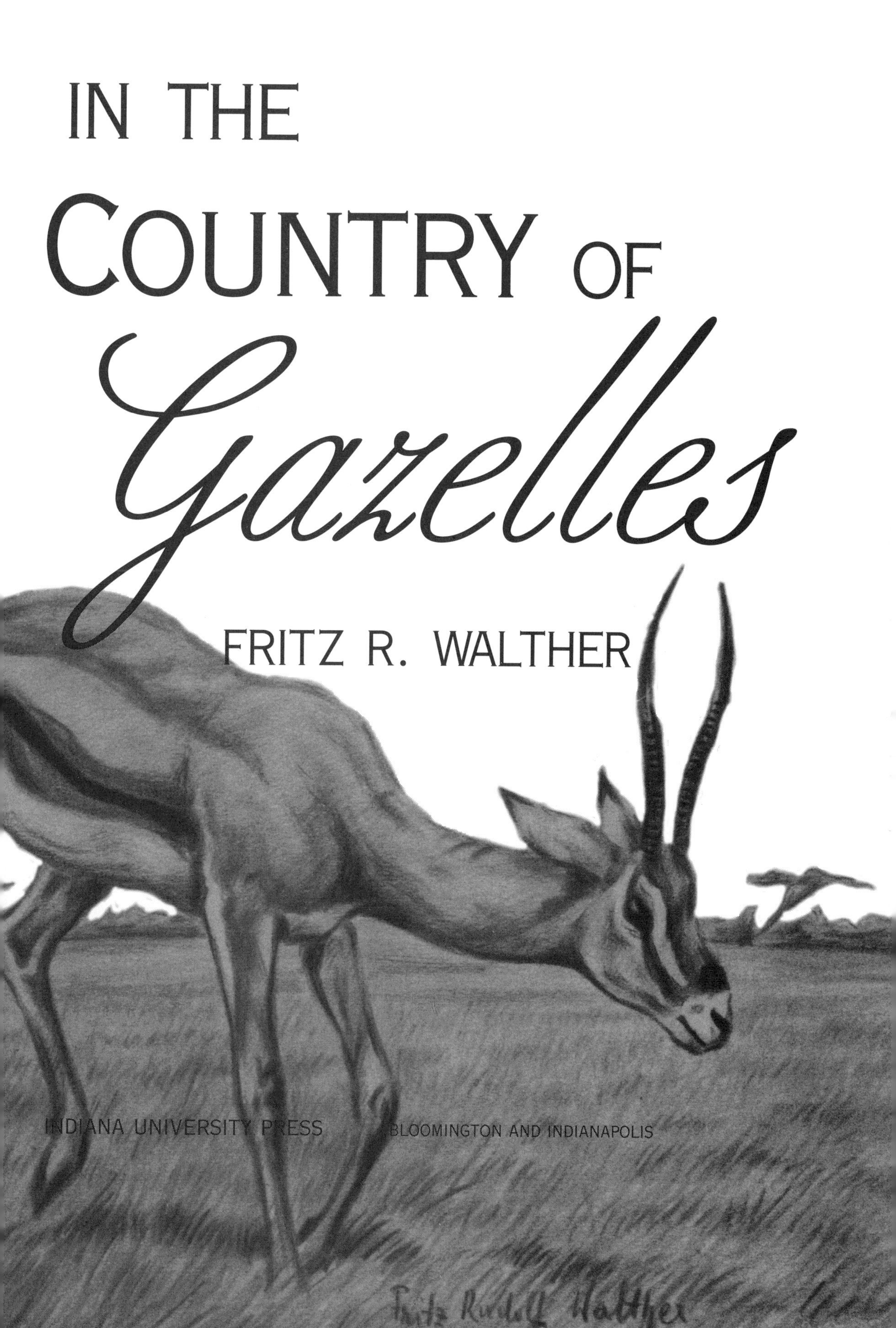

INDIANA UNIVERSITY PRESS BLOOMINGTON AND INDIANAPOLIS

The paper used in this publication meets the minimum requirements of American National Standard for Information Sciences—Permanence of Paper for Printed Library Materials, ANSI Z39.48-1984.

Manufactured in the United States of America

Library of Congress Cataloging-in-Publication Data

Walther, Fritz R.
In the country of gazelles / Fritz R. Walther.
p. cm.
Includes bibliographical references (p.) and index.
ISBN 0-253-36325-X (cl)
1. Mammals—Behavior—Tanzania—Serengeti National Park. 2. Birds—Behavior—Tanzania—Serengeti National Park. 3. Gazelles—Behavior—Tanzania—Serengeti National Park. 4. Serengeti National Park (Tanzania) I. Title.
QL731.T35W35 1995
599.73′58—dc20 94-21725

1 2 3 4 5 00 99 98 97 96 95

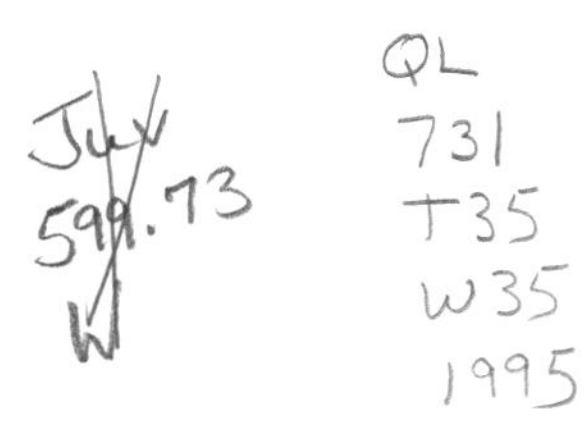
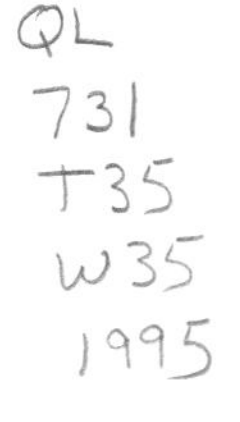

CONTENTS

Preface

When a man has devoted almost half of his life to something as "way out" as the study of the behavior of horned ungulates, he probably owes the reader some explanation as to how he came to do it. As impolite as it may be, I therefore would like to talk first about myself.

My keen interest in wild oxen, sheep, goats, and, above all, antelopes and gazelles, as well as my enthusiasm for Africa, go back to the days when I was a boy growing up in the city of Dresden, Germany. My grandfather, living in retirement, took me to the zoo there once a week when I was five or six years old. I cannot say precisely how it happened, but even at this early age, of all the animals—although I liked every one of them—I was most impressed with the antelopes and gazelles, and when I asked my grandfather where these beautiful creatures came from, his answer was usually "from Africa." So Africa became the land of my dreams. At the age of eight, I read my first book about Africa, and thus was born my determination to become "a leader of expeditions in Africa" when I grew up. Some years later, one of the few excellent grades which ornamented my generally rather mediocre school records was in geography—of course, Africa was the subject of our lessons at the time.

When I finished school, I was still fond of the idea of going to Africa as soon as possible; however, a minor obstacle intervened: This was World War II. I was drafted within the first year of the war. On the day of the ceasefire, I was on the eastern front, and—like all the German soldiers there—I was taken as a "prisoner of war" by the Russians. I was released from captivity in 1948, and I returned "home." The trouble was that the city of Dresden had been badly destroyed during the war. I found lodging in the home of a true friend in the city of Wiesbaden, and I became a teacher at an elementary school there. Besides teaching, I studied zoology, psychology, and philosophy at the University of Frankfurt am Main. Because I could devote no more than half of my time to the academic training, it took me much longer than a normal student. But eventually I earned my Ph.D. In between, I had married and become the father of two children.

A new branch of science was emerging at the time—ethology, the study of species-specific behavior in animals. Although I had never taken a lecture on the subject—simply because none had been offered as yet at the university—I became very interested in it. Following my old devotion to antelopes and gazelles, I began to investigate the behavior of such animals in several zoological gardens—still in addition to teaching, and thus predominantly on Saturdays and Sundays and during school vacations. My first paper on the subject was published in 1958, and more soon followed. At that time, very little was known about the behavior of these animals, and I earned a reputation as a specialist in this field. In truth, I merely played the somewhat dubious part of the one-eyed man who becomes king of the blind. Anyway, on the basis of this reputation, I got a position at a private zoo at Kronberg in Germany in 1960.

During this time, my plans to go to Africa had been somewhat pushed into the background. Deep in my heart, I had never quite given up the hope of seeing my "promised land," although I had no idea how to realize it, not the least with respect to the financial part. Meanwhile, however, Dr. Bernhard Grzimek and Dr. Richard Faust, directors of the Zoological Garden of Frankfurt am Main, had become involved in wildlife conservation in East Africa. To increase tourism to the national parks there, which at the time was still fairly light, they had established contacts with several tourist agencies. One day in 1961, they offered me the opportunity to guide a photo safari to the East African national parks. So I came to Africa for the first time—as a cicerone, and only for three weeks. Nevertheless, my childhood dreams had come true, although somewhat differently and later than expected.

During the following three years, I also guided tourist trips to African national parks. After I had spent another year at the Zurich Zoo in Switzerland working under the famous zoo director and animal psychologist Dr. Heini Hediger, my "shining hour" came in 1965. I was awarded a grant to study the Thomson's and Grant's gazelles in the Serengeti National Park in Tanzania for two years. The big predators in this park are the major tourist attraction. At least seasonally and/or locally, the two gazelle species are their main dietary staple, and their welfare, if not their existence, appeared to depend on them to quite an extent. So the administration of the Serengeti National Park considered the gazelles to be very important prey species and wished to obtain more detailed information on their life habits and behavior. Thus, the park's interests and my own research interests coincided quite happily.

From 1958 through 1967, I had published one book and about

thirty papers on the behavior of "my" animals, and in fall 1967 I got a call from the United States to join the Department of Zoology at the University of Missouri, Columbia, where I worked for three years. I then accepted an offer from Texas A&M University, College Station, where I was with the Department of Wildlife and Fisheries Sciences for fourteen years. While still based in Texas, I spent another year in the Serengeti National Park in East Africa and several months in the Kruger National Park in South Africa and in the Etosha National Park in Southwest Africa, in addition to taking shorter trips to other countries.

In the course of those years, I became an old man. I retired from Texas A&M and returned to my homeland. I am not sure whether this was a wise decision, but "an old hare returns to the area where he was born," as the hunters say. (Since I never did research on hares, I cannot tell from my own experience whether it is true.) Now I live with my wife and my dogs in the Western Forest in Germany. My days of field research overseas are over, and only my memories return to "the country of gazelles," where I had the greatest time of my life. As is well known, a man has nothing to do in retirement, and thus I sometimes would sit down and write up some observations, experiences, and thoughts from those "good old days." I also completed and elaborated some of my earlier sketches, and I put together a few series of photos. This book is the result.

IN THE COUNTRY OF *Gazelles*

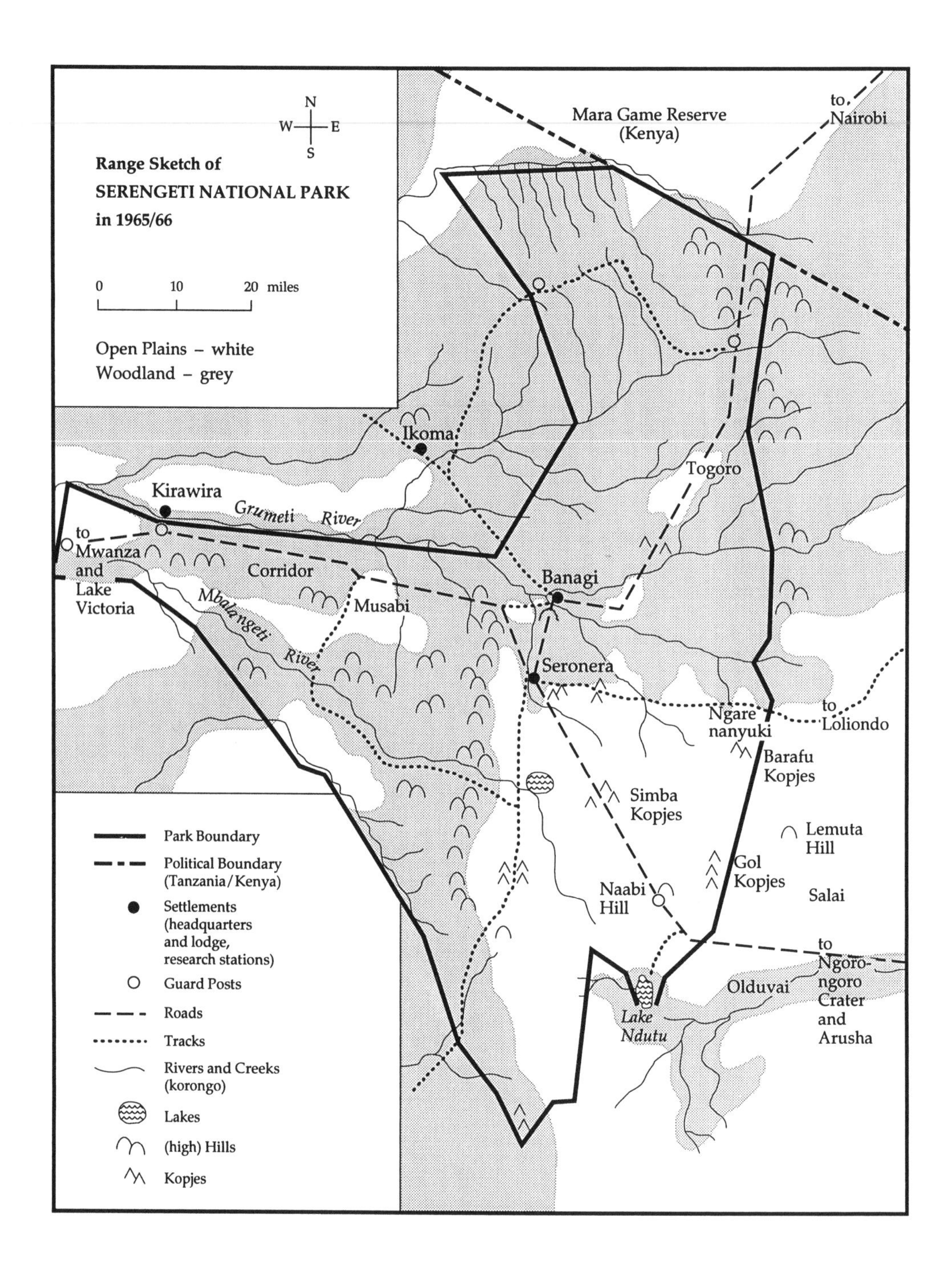

N
W
E
S
Range Sketch of
SERENGETI NATIONAL PARK
in 1965/66
0
10
20 miles
Open Plains – white
Woodland – grey
Mara Game Reserve
(Kenya)
to
Nairobi
Ikoma
Togoro
Kirawira
Grumeti River
to
Mwanza
and
Lake
Victoria
Corridor
Musabi
Banagi
Mbalangeti River
Seronera
to
Loliondo
Ngare
nanyuki
Barafu
Kopjes
Simba
Kopjes
Lemuta
Hill
Gol
Kopjes
Naabi
Hill
Salai
to
Ngoro-
ngoro
Crater
and
Arusha
Olduvai
Lake
Ndutu
Park Boundary
Political Boundary
(Tanzania/Kenya)
Settlements
(headquarters
and lodge,
research stations)
Guard Posts
Roads
Tracks
Rivers and Creeks
(korongo)
Lakes
(high) Hills
Kopjes

1 Short-Tail and Roman

On January 1, 1965, I arrived in East Africa to conduct a two-year study on the behavior of Thomson's and Grant's gazelles in the Serengeti National Park. Virtually nothing had been published on the behavior of Grant's gazelle at that time. On Thomson's gazelle, only the British ecologist Brooks had published a booklet, in which he also provided some information on the behavior of this species. Among other things, he spoke of territorial males, individuals who occupied and maintained a specific place, a territory, in their own right. He expressed the opinion that these were males who were without an established harem (a group of females which remain together with a particular male and in which he holds the exclusive right of copulation), and thus, males who either had not succeeded in forming such a harem, or had possessed one but later lost it to another male. In short, one could conclude from his description that the territorial bucks were the less successful, the overly weak, old, or young ones. During short-term visits to East African national parks in the four previous years, particularly during a four-week stay in the Ngorongoro Crater on the kind invitation of the American ethologist Richard D. Estes, however, I had learned that this view could not be correct but that territoriality apparently played a very important role in the life history of Thomson's gazelle. So one of my first tasks was now to take a closer look at it.

Two problems arose in this context: For one, I would have preferred, at least in the beginning, to concentrate my research on a limited observation area where a few gazelles would be present at any time throughout the seasons. Such an area is not easily found in the vast Serengeti plains, in which the herds are frequently on the move. Secondly, I had to learn to recognize a few territorial bucks individually, since when an observer has seen a particular male at a particular place and returns to the same area some time later and again meets a buck there, it is extremely important with respect to the question of territorial behavior to be able to ascertain whether this is the same male as before, or another one.

Of course, it is possible to artificially mark a number of individual animals, but I have always hated to do so, although I have to admit that for certain purposes—which, however, in my opinion, are much rarer than many of my colleagues apparently think—it hardly can be avoided. In order for the animals to be marked, they first have to be captured or immobilized with a capture gun. This always is an expensive and time-consuming job, and it sometimes may also jeopardize the marked animals' lives, particularly when they are as small and delicate as Thomson's gazelles. Also, more than once a scientist has ended his stay in the field only to tell an admiring audience how many animals he captured and marked but that the time available had not been sufficient for any further research.

Furthermore, we usually do not know exactly whether and to what extent we may shorten the life span of artificially marked animals, especially when they have many enemies, as is true for gazelles. Sometimes the marks may be striking not only to the scientific observer but to predators and poachers as well, and may stimulate them to hunt just for the marked individuals.

Finally, it simply is against all I stand for to "ornament" free-ranging wild animals with collars, ear clips, ties, and similar products of civilization, particularly in one of the few areas on earth where the land and the creatures are still relatively free of human control and interference. Thus, I hoped that I might be able to avoid the marking business and everything associated with it.

I asked the team leader of the research station of Banagi for advice. At that time, this was the British ecologist Dr. Phil Glover. Later I could

In the Togoro plains during rainy season.

proudly speak of him as my friend. He was old and wise enough not to tell another researcher what to do or not to do—as is the mistake of so many people in comparable positions. But he knew the country extremely well, he was readily available when anybody needed his help, and his advice was always as good as gold. He suggested that I go to the Togoro plains, which were only some twenty miles north of Banagi and where, in his experience, "tommies"—as Thomson's gazelle are usually called in East Africa—could be seen at any season.

In 1965/66 only a small "road" led from Banagi to the Togoro plains (and farther north). It sometimes was a difficult route during the rainy season, and the density of the "traffic" was about two to three cars per week. Only ten years later, this road was four times as wide, and every day twenty to thirty cars drove along it to the tourist lodges Lobo and Keekorok, which did not exist as yet in my day. (Actually, the construction of Keekorok in Kenya had just started at that time.) To keep this road in good shape, a big gravel pit has now been established in Togoro, with roads leading to and from it. Also, several enclosures surrounded by wire fences have been installed there for the purpose of ecological research. The researchers have found out this way that the grass grows longer when and where the animals do not eat it. (As my Canadian friend and colleague Val Geist once put it: Ecology is the painful elaboration of the obvious.) In short, nowadays Togoro is opened up for tourism and science, as the progressionists like to phrase it in their shark language. Only one thing has gone this way: Togoro.

Thanks to Phil Glover's advice, I had found a good observation area. However, he could not tell me how to recognize individual tommy males and to distinguish them from each other. In this regard, I trusted my good luck and my experience obtained with captive animals in zoological gardens. After all, a good game warden in the United States or a good forester in Germany will usually know some of the deer in his district individually without having marked them.

As I found out later, most tommies have a black spot on the nose and a white spot on the forehead, which vary individually (and probably also with age) as to size and shape in this species.* So a limited number of bucks—I eventually came to know fifty-four in Togoro—can be individually recognized and quite reliably distinguished by these nose and forehead spots, particularly when they stay within their relatively small territories. When I later measured the size of seventeen tommy territories, I found the smallest to cover not quite two acres, the largest about twenty-five acres, and the majority from somewhat more than two up to twelve acres.

At first, of course, I was not aware of the variation in nose and forehead spots. However, I soon learned to individually distinguish two obviously territorial tommy bucks physiognomically from all the other males in Togoro. One of them, with relatively short and V-shaped diverging horns, attracted my attention because he had only half his tail. This sometimes may happen in Thomson's gazelle; occasionally one will have barely any tail left. Possibly they bite their tails as young animals when they are taken ill by sarcoptic mange (the only disease that occurs with some frequency in Thomson's gazelle in the Serengeti area), since this

*In Grant's gazelle, this is not the case. Grant's gazelle have a black spot on the nose, too, but it is much more uniform in size and shape from one individual to the next than in the tommy. The shape of the horns is more variable in Grant's males than in tommy bucks, however, and thus, to some extent, Grant's bucks may be differentiated by their horns.

mange often begins in the tail region. The other territorial tommy males in the area had their complete tails. Thus, the one with half a tail was easily recognized. Nobody is totally free of vanity. When guests would come to visit me in Banagi at that time, I liked to impress them a little by saying: "We will now drive about twenty miles north. Then we will leave the road and turn to the left. About six hundred yards off the road, there is a single small acacia tree. Near it or possibly even right below it you will see a tommy buck, and this tommy buck will have only half a tail." It always turned out to be true.

One of the neighbors of this "Short-tail" had long, almost parallel standing horns, was slightly darker in coat color than most of the other bucks, and seemed extraordinarily at ease with my car and me from the very beginning. I cannot explain it, but he was and remained the only gazelle in Serengeti which I could approach on foot up to a distance of five yards. With the Land Rover, a four-wheel-drive, off-road vehicle commonly used in field work in East Africa, I sometimes could get even closer. Generally, the game animals in the Serengeti National Park are considerably shyer toward people than toward cars. They are not used to the sight of pedestrians since tourists are not allowed to leave their cars, and the game wardens and scientists also do most of their work from their cars. But even by car, I could never get any closer to Short-tail than twenty-five yards in the best of circumstances.

Scientists often call observed animals by letters and numbers; thus they may speak of the male "M 12" or the female "F 21." I always gave names to "my" bucks. This does not sound as scientific and objective, but it has the great advantage that, when looking at my notes even after years, I could always remember immediately who was meant. I named Short-tail's long-horned neighbor "Old Roman" or just "Roman." Somewhere I had once read that the Roman legionnaires had the letters SI on their helmets,* and that these were abbreviations for the Latin words *semper idem* (always the same). As I soon found out, it was "always the same" buck who stood there in his territory in sunshine and in rain, during the hot days and the cool nights, with or without females. Thus, he deserved this honorable name. Certainly, it would have fit a number of other territorial males, too; however, to me "Old Roman" became the prototype of a good territorial tommy buck.

The first remarkable event I observed involving Short-tail and Roman

*Years later, I was told that my information probably had been historically incorrect, that some soldiers had had these letters on their helmets, but not the Roman legionnaires.

Tommy buck "Old Roman" in Togoro.

was a big fight. It was during that particularly fine time when many things in Serengeti were still relatively new to me and every day brought another "first." I had been in Togoro for the first time, had determined the whereabouts of several tommy territories, had made the acquaintance of Short-tail and Roman, and finally, with the sinking sun, I had found a place in their neighborhood where I intended to stay and spend the night in my car. Suddenly, grayish-white clouds of dust rose in front of the orange evening sky, and with my binoculars I watched as the two fighters engaged in a mighty struggle.

In flat jumps with all four legs, Short-tail and Roman clashed together, "deep diving" with their heads, spreading their forelegs, and kicking their hindlegs high up into the air. Then both jumped back simultaneously and immediately forward again for another clash. Sometimes they became locked together with crossed horns, and then one

A fighter pushes back his opponent with a powerful forward-thrust attack. The defender jumps backward (Thomson's gazelle).

pushed the other back over ten to twenty yards, only to be moved back the same distance himself in the next moment. After twenty minutes of vehement fighting, the two were somewhat out of breath. Simultaneously they began to graze and turned away from each other.

Later, when I knew more about the behavior of gazelles, it became clear to me that these two males probably had established their territories only a short time previously. In Thomson's gazelle, fights between territorial neighbors are, so to speak, demanded by good breeding; however, they usually are not so vehement and, above all, do not last as long. Long and strenuous fights, in which one of the two combatants may sometimes be vanquished and chased away, occur predominantly within the first days or even the first hours of establishing territories. Furthermore, many tommy males establish their territories at the beginning of a rainy season (although in this species, territoriality is not necessarily limited to a specific season). In 1964/65, the "small rains" had not started according to the calendar, i.e., in October, but waited until the end of December 1964. The aforementioned fight took place in the very first days of January 1965. Thus, this point of time also supports the view that the two males probably had established their territories only a short while before.

Since Roman was so at ease with me, I was able to extensively study the daily activity of a territorial tommy buck through him. As far as he was concerned, I could move by car inside his territory and accompany him on his daily rounds through it. When I wished to learn something about his behavior toward other gazelles, however, I had to stay at the boundary of his territory because the others were not as comfortable with my presence, and especially the females—who in general proved to be shyer than the males—would not enter his territory when I was standing in the middle of it.

Later I learned how to observe almost all the gazelles (as well as other open plains animals) from a rather short distance. It is not difficult, but you have to know how to do it. The observer drives into a good game area and then remains in one spot. Three points are important: The car must not move from its place during the period of observation—in my case, this usually meant two to three days. The observer may leave his car only very carefully and for a very short time, as far as cannot be avoided for bodily needs; otherwise he must remain inside the car at all times. Finally, the car has to be parked close to a somewhat prominent natural object such as a tree, a bush, a rock, or a termite mound. If the car is the single conspicuous object within the vast surroundings, the game animals may keep a distance of one hundred to two hundred yards or even more even after several days. However, when the above three points are observed, the animals will "forget" the presence of the car within a day or even a few hours, and they will approach it to within about fifteen yards, occasionally even closer.

As noted, the game animals in the Serengeti National Park are more familiar with cars than with people, which is expressed by their flight distance from them. On this basis, I conducted a few little "experiments." When I succeeded in leaving my car without the gazelles becoming aware of it, and then stood motionless beside it, I could stand in full sight of the animals at distances of only fifteen to thirty yards away without causing any reaction. Like many other game species, gazelles can see silhouettes and motion very well, but they have difficulty discriminating between objects placed closely behind each other. As long as my silhouette remained inside the silhouette of the car, they did not recognize the human figure. However, when I managed—again without the gazelles becoming aware of it—to place myself a few steps away from the car so that they now could see the vehicle and the man as two separate figures, they fled immediately over one hundred to three hundred yards (except, of course, Old Roman), even when I remained absolutely motionless.

Another interesting discovery, which apparently is valid only for Thomson's gazelle but not for Grant's gazelle and other game species, I made by mere chance. A scientist alone in the plains for several days has to pay attention so as not to lose track of the dates. Therefore, the first thing I did every morning was to glance at the little pocket calendar which my wife had given me as a present at our last Christmas Eve together, to scratch out the date of the passed day and transfer the new date into my diary. One morning I woke up in my car, as usual surrounded by tommies, and I read in my calendar the words "Mardi gras" below the date of the new day. Remembering the masquerades and parades and all the joking going on at home on this day, I felt a little homesick. Among the supplies in my car, I had a small bottle of rum—in case of an emergency. I decided to empty this bottle in order to celebrate Mardi gras. Rum on an empty stomach makes a man feel very cheerful. So I started singing the most popular Mardi gras song in Germany: "Sooo ein Tag, so wunderschön wie heute, so ein Tag, der dürfte nie vergehn" (Suuuch a day as beautiful as this day, such a day should never end). My singing was not very beautiful but it was loud, and since I could not remember the other verses, I repeated the first one fifteen times. Suddenly, I realized that all the gazelles were still around my car, grazing peacefully despite my loud singing. Indeed, I found Thomson's gazelle to be strikingly indifferent to many artificial noises, although I am almost sure that they can hear them. For instance, they will promptly react to soft vocalizations on the part of their conspecifics. At least in a national park, however, a man can speak loudly in their presence and even bang the door of his car. If they do not see the movement, they do not react. I know of only one exception: They regularly run upon hearing the mean screaming caused by a thorn scratching along the body of a slowly driving car. They also do not care very much about natural sounds other than from conspecifics. For example, they hardly react to the shrill alarm-whistling of reedbucks, and even the roaring of lions does not impress them. When I was a boy, I read in a book that all prey animals would tremble and run as soon as they heard the voice of "the king of the beasts." I don't know who wrote that book, but I am sure of one thing: the author knew nothing about gazelles.

Twenty-four-hour observations are matters of their own. Up until midnight, staying awake was fun for me, and up until three in the morning, it was still relatively easy. However, during the following hours, I had a tough struggle against falling asleep. In the old days, sailors spoke about the hours of the "dogwatch." They knew what they were talking about. With the first morning light, however, all my exhaustion was gone. There is hardly any greater experience than daybreak in the free African plains—and it is a true grace to be alone in such hours.

When the night changes to dawn, when the silhouettes of the game animals show up like shadows in the surroundings, when the first voices of birds are heard, when the wind freshens up, when the eastern sky flames up red, when the red changes into a pale yellow, and when finally the sun breaks through at the horizon—"The heavens praise the Lord." In such hours I have folded my hands in prayer, a prayer of gratitude for the overwhelming beauty of God's creation and that I was among the few people privileged to still be able to experience it in modern, nature-destroying times.

The days when the kori bustards were mating were quite amusing. These big birds are common in the East African plains, and the gazelles are used to them. Often they will meet them at close range. In the prephase of mating, the male kori bustard erects his tail and his somewhat ruffled neck vertically. This "basket-like" posture may occur at any time of the day and is maintained for hours. It is familiar to the gazelles, too, and they do not react to it. However, when the display reaches its culmination point, which frequently happens in the early morning, the kori cock erects his whole body and ruffles the white feathers of his neck so that the neck appears to be as thick as the body. His beak and the black cap on his head resemble a lid on a bottle. Thus he no longer looks like a bird but like a strange post standing amid the plains. From time to time, the bustard will throw up the feathers of his neck in a jerky fashion, such that two dark spots the size of a man's fist become visible at his nape. Simultaneously, a sound is heard similar to the muffled beat of a big drum.

For the younger gazelles, this is all quite amazing. They approach from all sides, watching the strange spectacle with their necks highly erect. Apparently they can still recognize the front and rear of the bird, since they eventually will gather behind it. Then a particularly bold and curious youngster may approach very closely, stretch his head and neck forward, and touch the white underside of the bustard's erect tail with his nose. This is too much for the big bird. He interrupts his display, spreads his wings, and turns hissing toward the mischief-maker. In doing so, the bird resumes the usual shape well known to the young gazelles. They step aside and calm down—until the bustard begins to display again, attracting their attention and curiosity once more.

Of course, an old territorial tommy buck such as Roman is above such infantile stupidities. To him, a bustard, even in maximum display, is just a common, ordinary sight, not worth paying any attention to. Other things are much more important to him.

Roman has spent the last hours before daybreak (about 6:00 A.M.)

When the display of a kori bustard reaches its culmination point, curious tommy youngsters may approach the cock from the rear.

alone resting in his territory—lucky guy, he doesn't have to stay awake during the hours of the "dogwatch." He rises with the first light and takes a "marking walk" along the boundary of his territory, freshening up the dung piles as well as the secretion marks which he has deposited from his preorbital glands onto single grass stems—not all of them, of course, but quite a number.

There are ten to twenty dung piles along the boundary of a well-established tommy territory, and one or two approximately in the center. Roman walks to one of these dung piles and scrapes the ground there with a foreleg several times, sometimes also once more with the other foreleg. He then steps forward with both forelegs, keeping his hindlegs in place. He stretches his body and lowers his belly toward the dung pile in this way, looking somewhat like a gymnast doing a "push-up." He urinates in this posture. Immediately after urination, he abruptly brings his hindlegs forward so that their hooves are placed almost laterally to the stationary forelegs—moving from the "push-up" posture into a squatting posture. With his back crouched, his hindquarter is now not much more than a hand's breadth above the ground. He places his droppings on the urine, both right on top of the (already existing) dung pile. He then walks to the next dung pile to repeat the procedure, and so he may attend several dung piles, one after the other.

Between the dung piles, there are chains of the secretion marks which a tommy male places at the tips of relatively high grass stems,

Urination and defecation follow each other in a sequence, resulting in the establishment of dung piles by territorial tommy bucks.

preferably of his own body height. Roman freshens up some of these marks and adds a few new ones. He stops walking, sniffs at a grass stem, twists his head so that one cheek points more or less to the ground, and opens his preorbital glands. He carefully lowers the preorbital region toward the grass, bringing the stem's tip into the widely opened gland.

In Thomson's gazelle, the preorbital gland is a small "bag" at the corner of the eye (where humans have the lachrymal gland—which otherwise is quite different from the preorbital gland). This "bag" has an opening in the form of a vertical slit which is closed in a state of rest but which the male can open at will. Inside the gland-"bag" there is a brownish-blackish grease, the secretion. When marking the tip of a grass stem, the tommy buck deposits some of this substance with quivering movements of the opened gland's margin. The secretion soon hardens in the open air. When a male marks the same grass stem several times, eventually a pea-sized secretion pearl may stick at its tip, sparkling in the sun.

People with obviously a keener sense of smell than I have say that there is a weak aromatic scent to this secretion. As far as can be concluded from the behavior of the gazelles, they are able to perceive a scent only when they almost touch the secretion mark by nose. However, they probably can see the described secretion pearls from a somewhat greater distance.

Above: Tommy buck marking a grass stem with secretion from his preorbital gland. *Below:* The tip of the marked grass stem is covered with a dark secretion.

Certain people think the dung piles and secretion marks may have an intimidating effect upon other males, preventing them from entering a territory. I have seen only a very few incidents which possibly could be interpreted this way. On the other hand, it happens every day that "bachelors" (nonterritorial males) will enter a well-marked tommy territory, paying no attention to the dung piles and secretion marks of the owner. Thus, if there is any intimidating effect, it certainly is weak and unreliable. From the evidence it seems more probable that the dung piles and secretion marks make it easier for territorial neighbors to recognize the boundaries of the adjacent territories. Most likely, however, they serve the better orientation of the owner of the territory himself.

By the way, nonterritorial tommy males also mark with their preorbital glands, even during migration, and they urinate and defecate very similarly to the owners of territories, although not as frequently and usually not several times at the same spot. Thus, these behavior patterns and the marks, respectively, indicate the presence or the passing through of tommy bucks. The marking of a territory is only one, albeit probably quite important, special case in which a single buck individually marks a limited area.

During a "marking walk," a territorial male occasionally may gore the grass, the soil, or a bush with his horns. These acts of aggression against inanimate objects of the environment, as well as the depositing of preorbital gland secretion and the urination-defecation sequence, may occur at irregular intervals throughout the day. Moreover, they frequently show up on special occasions, for instance, in connection with aggressive interactions.

After his morning "marking walk," Roman starts grazing, often near the boundary, a situation which may easily lead to conflict with the corresponding neighbor. When no females have entered his territory by eight or nine o'clock, he beds down in the open with his back to the east, toward the sun and the wind. Perhaps some females have already arrived in the territory of his neighbor Short-tail, who then remains on his feet. Otherwise, this morning rest with back toward and face away from the sun is obligatory in all the gazelles. They are warming up after the cool hours of the early morning.

Between ten and eleven A.M., they become active again. Now some of the females, who have been walking and grazing in Short-tail's territory, may move into Roman's territory. As soon as they pass the boundary, Roman approaches them—always one female at a time—with his neck and head stretched forward, with widely opened preorbital glands, and with soft "bl-bl-bl"-vocalizations uttered through his nose. He

Tommy buck goring the grass (object aggression).

places himself behind a female and herds her toward the center of his territory. In doing so, he erects his neck steeply, lifting his nose toward the sky, and in walking he beats a kind of "drumroll" with his stiffly stretched forelegs. When he has succeeded in driving a few females ahead, the others soon follow, and finally, there stand and graze usually twenty to thirty but sometimes a hundred or more females inside his territory.

Again Roman approaches some of the females. When a male follows a female persistently enough, she almost regularly urinates. The tommy buck sniffs at her urine on the ground, raises his head, and opens his mouth slightly. He remains in this posture, called Flehmen, for a little while, then closes his mouth and briefly licks his upper lip. Obviously, a male can recognize whether a female is in heat by sniffing at her urine and subsequent Flehmen, because after Flehmen, either he loses all interest in this female, or he begins to court her intensely. Frequently a tommy buck will "test" several females, one after the other, sometimes even the whole herd which has arrived in his territory.

How about that—Roman has found a female in the "right" state.

When a female, driven by the male, urinates, the tommy buck sniffs at her urine on the ground and performs Flehmen, a behavior by which he apparently can recognize whether a female is in heat.

While he stood performing Flehmen, she walked on. Now he follows her quickly with widely opened preorbital glands and drives her, showing neck-stretch (head and neck stretched forward) and nose-up (neck erect, nose more or less vertically pointing toward the sky) displays alternately. A few times, the female turns sharply with a jump and moves off, passing Roman in the opposite direction—apparently a maneuver to get rid of the driving male. However, he follows her persistently and starts—now in normal posture—to treat her with foreleg kicks. He raises his stretched foreleg abruptly up to 45° or even almost 90°, sometimes when he is still one or two yards away from her, but quite often when he is so close behind her that his kicks may go between her hindlegs or laterally along the outside of them. Particularly when a female stops walking or even starts grazing, the male will deliver one foreleg kick after another until she walks again and the mating march is continued.

Suddenly Roman stops and looks up. A bachelor, a nonterritorial

MALE COURTSHIP DISPLAYS IN THOMSON'S GAZELLE

Above: The tommy buck approaches the female, stretching his neck and head forward (neck-stretch). *Center:* Nose-up posture in close approach. *Below:* Foreleg kick during mating march.

In "en garde" posture with highly presented horns, a territorial tommy buck (Roman) approaches an intruder, who scratches his neck before he takes to flight.

buck, has entered his territory and is moving quite happily and frankly straight toward the females in it. "Politics before love." Roman leaves his favorite in her place, erects his neck, and pulls his chin toward his throat so that his horns, presented to the other male, "tower" above his head. In this "en garde" posture he walks with widely opened preorbital glands toward the intruder. Becoming aware of Roman's approach, the bachelor stops moving, lowers his head to the right and the left as if he were going to graze, lifts his feet alternately as if the ground below him suddenly had become hot, grooms his flank with his mouth as if insects were bothering him there, vigorously scratches his neck with a hindleg, shakes his flank, and finally flees at a gallop.

Roman returns to his chosen female and starts the mating ritual from the beginning. Eventually the female walks at a steady but quiet pace, Roman without any special display directly behind her. The pair now moves in a zigzag course at the periphery of Roman's territory. Immediately on the other side of the boundary, Short-tail grazes in his territory.

In mounting, the gazelle buck walks bipedally behind the female.

Finally, the female stretches her tail horizontally, in Thomson's gazelle the indication that she is ready for copulation. Roman mounts her while walking. He rises on his hindlegs, angles his forelegs toward his chest, and walks bipedally behind her without touching her with his body or forelegs. Now one mount follows the other, and finally he copulates with his body so steeply erect that for a moment it looks as if he might fall on his back. (Very exceptionally, this literally can happen to a gazelle male.) After copulation, all the mating activity is over. Both partners stand quietly and relaxed, then walk away from each other in different directions.

Meanwhile the female herd has crossed Roman's territory and arrived at the opposite boundary. A few females have already crossed the boundary and moved into the neighboring territory. Roman hurries

In territorial herding, a tommy buck passes a running female in order to cut her off and block her path.

toward them and tries to herd at least some of them back, but he is not very successful. Eventually only one female is still with him, but she, too, tries hard to leave his territory and follow the others. Roman chases her back toward the center, but she suddenly turns and runs again toward the boundary at full gallop. Roman pursues her as fast as he can. He passes her and turns into broadside position in front of her to block her path. The female stops, turns, walks back a few steps, and again tries to break through at a gallop. Roman manages to stop her a few more times, but finally she crosses the boundary. Roman blindly follows her in a wild chase and runs far into his neighbor's territory. The neighbor then charges him and they clash together, both from full gallop. After several violent clashes, Roman retreats to his territory.

Meanwhile, more females, stragglers of the herd which passed here before, have come from the other side and arrived in neighbor Short-tail's territory. He has found one of them to be in heat and is now courting her intensely. However, this female is not very cooperative. Instead of walking in front of the male, she trots and gallops and makes one sharp turn after the other. Eventually Short-tail "loses patience" and chases her violently—almost as if she were a fleeing rival. Roman stands highly erect at the boundary and watches the wild chase. The female runs into his territory, and Short-tail stops right at the boundary in a cloud of dust. Roman takes over and chases this female across his territory. This time he respects the (opposite) boundary and stops there from full gallop. The neighbor on this side takes over and chases the female. Breathing with open mouth and obviously exhausted, she now slows down to a walk. Immediately this male starts courting and driving her. After a short mating march, she stretches her tail horizontally. The buck copulates after several mounts.

It is now high noon, and the sun is burning vertically from the sky. Roman is alone again in his territory. A little shade would be appreciated, but he has no tree inside his territory. However, there is a big acacia tree only a few yards beyond the boundary in a neighboring territory. Roman walks to this tree and stands unhindered in its shade during the hottest hour of the day. During this heat (35° C in the shade), nobody is too punctilious. This neighbor has another, though very small, tree in the center of his territory and stands there. The tommies are true artists in taking advantage of the slightest shade. Not that they really need it—there are many tommy territories in areas with no trees at all. Also, Roman did not go to this tree every day. When he did go, he always returned to his territory within about an hour.

Roman passes the afternoon grazing and marking. About five o'clock P.M., he beds down for half an hour. Then he grazes again. The females now return from the direction in which they had disappeared at noon. Upon their return, the herding, testing, and courting activities of the territorial bucks start again. However, this time the females have come from an area where there are no territories. There they have intermingled with a bachelor herd, and quite a number of the nonterritorial males are still with them. Roman makes short work of them. Without any special display posture but with loud vocalizations—they sound like "pshorre-pshorre-pshorre" (o as in "on," e as in "echo") and are uttered by nose—he rushes toward the bachelor next to him. That is all it takes. Only exceptionally will a bachelor fight back when charged by a territorial buck. Normally, he will flee as fast as he can upon hearing the described "chasing call." Having chased the first bachelor out of his territory, Roman charges the next one, and so on. His neighbors are also chasing the nonterritorial males, and soon the female herd is "cleaned out."

As the sun sinks, the females have moved ahead and left the territories. Now Roman has an encounter with Short-tail. They meet right at the

The tommies are artists at taking advantage of the smallest shade.

boundary, both in "en garde" posture with highly presented horns. They flatly jump toward each other and clash their horns together near the ground, then they jump backward and immediately forward again for another clash. After several such confrontations, both start grazing simultaneously, still facing each other. While grazing, they step backward, thereby increasing the distance between them. From this position they can charge again at any time. When this does not happen, they turn into a lateral position—either parallel or reverse-parallel (head-to-tail)—and graze uninterruptedly. In parallel grazing, they may move along the entire mutual boundary of their territories and back again. They can still go back to

A clash with full force (Thomson's gazelle).

fighting from this lateral position, but this happens much less frequently than from frontal grazing. Eventually they turn their hindquarters toward each other. Each of them now moves toward the center of his territory, occasionally interrupting his grazing to mark a grass stem with preorbital gland secretion and/or to urinate and defecate in striking postures.

In some such boundary encounters, the territorial neighbors do not actually clash together but only perform an "air-cushion" fight. They rush toward each other, jumping back and forth and showing all the maneuvers of attack and defense as in a true fight, except that they do not touch each other. It is as if there were an invisible cushion between the horns of the two fighters. In every case, however, the "grazing ritual" described above ends an encounter between territorial tommy bucks—the two first graze in frontal position, later in parallel or reverse-parallel position, and finally they turn completely around and move away from each other while constantly grazing. Sometimes a grazing ritual will take place even without any prior fighting. There is no winner or loser in these encounters. The neighbors merely "ratify" the position of the boundary between their territories. Even when there are no females anywhere near, and the tommy bucks are completely alone in their territories, it frequently happens, particularly in the early evening hours, that one of them will walk to the boundary, place himself there in "en garde" posture with highly presented horns, and wait for his neighbor to become aware of him and his challenge, respond to it, and approach for a fight and a grazing ritual.

GRAZING RITUAL IN THOMSON'S GAZELLE

At the end of a boundary encounter, territorial tommy bucks (Roman and Kimm) change from fighting to grazing.

In frontal grazing, they step backward and increase the distance between them.

While grazing, they turn into broadside position.

Grazing continually, the opponents turn away from each other. Each of them moves back toward the center of his territory.

Air-cushion fight (Thomson's gazelle).

The herding of females, the chasing of bachelors, and the encounters among territorial neighbors are continued during the short tropical dusk. After nightfall (about 7:00 P.M.), the gazelles remain active for about half an hour. Then they bed down to rest. About ten o'clock they usually are on their feet, but they soon bed down again. There is a more pronounced activity period which begins shortly before midnight. On dark nights, the gazelles move and graze for about an hour. On nights with bright moonlight, however, this "midnight activity" can be continued up to four hours, and the territorial bucks mark, threaten, fight, chase, herd, and court as during daytime.

In Togoro, toward the end of this activity period, the females have usually left the area with the territorial males. Roman is alone in his territory. By about four o'clock in the morning, he—like all the other gazelles—beds down and rests until daybreak. Incidentally, gazelles spend most of the time during rest periods dozing and ruminating, but they sleep only for minutes at infrequent intervals. Five to ten minutes of uninterrupted sleep is a long sleeping period for an adult gazelle. (Young gazelles may sleep longer and more frequently.) All the short sleeping periods in a twenty-four-hour day added together may total one to two hours, whereas these animals easily spend eight to ten hours lying down.

This is a typical day for a tommy buck at the high point of his territorial period. Of course, there are variations, and sometimes there are special events. One day, I had parked my car as usual at the boundary of Roman's territory. His territory, as well as all the other territories around, was well frequented by females, and the territorial bucks were

Sometimes a tommy buck walks to the boundary of his territory, places himself there in "en garde" posture, and waits for his neighbor to become aware of him, take the challenge, and approach for a fight and a grazing ritual.

herding and courting them everywhere. There was also a bachelor herd not far away. Suddenly, all the females and bachelors began to run at once, fleeing at a full gallop for about half a mile. Then they stopped and looked in the direction from which they had come. Only the territorial males remained in their territories. What had happened?

Three cheetahs had shown up suddenly—and the tommy buck next to them was Roman. Highly erect, with vividly wagging tail, he stood in his territory and watched them. The cheetahs approached him to a distance of about one hundred fifty yards, sat down, and looked at him. Prey and predators faced each other. This is a typical situation from which a hunt can start out at any moment. As I had experienced on several occasions, cheetahs do not necessarily attack the prey animal next to them, particularly not when it is still standing. I sometimes even got the impression that they prefer to hunt fleeing prey. However, such a rule always has its exceptions, and it was by no means impossible that Roman would turn and flee eventually and release the cheetahs' pursuit in this way.

In principle, a good scientist should only observe "the course of nature" in such a case, and not interfere. Also, I had nothing against the cheetahs having a good breakfast—but not Roman! Objectively, my attitude was justifiable because he was one of the individual tommy males whom I knew best, whose life history I had followed for months, and which I wished to keep following as long as possible. However, I will readily admit that my subjective attachment to him played a role, too. Homo sum. . . .*

I started my car, put it quickly into first, second, and third gear, and rushed toward the cheetahs, scaring them away and spoiling their hunt for this time. By the way, my succor had an unexpected side effect which I would not dare mention had it not been so very striking. After this event, the gazelles in this area became more comfortable with my car. For instance, the females and bachelors would now come into Roman's territory even when my car stood right in the middle of it. They had never done this before.

Another time, a medical doctor from a hospital in the city of Mwanza near Lake Victoria visited me with his two sons for a couple of days. They were passionate photographers, but they did not have telephoto lenses for their cameras. In the Serengeti National Park, particularly in the area around the tourist lodge at Seronera, the animals are used to car traffic. Thus, it is not difficult to take good pictures of lions, buffalo, giraffes, etc. there, even without a telephoto lens. Gazelles, however, are usually somewhat shyer, and so the last day of my guests' vacation had come, and they had not been able to take a satisfactory picture of a gazelle. On the other hand, I had told them so much about "my people," the Thomson's gazelles, that they wanted to have a few good photos of them.

Since my guests were very nice persons, I wanted to do them a favor, and I took them with me to Togoro. While driving, I told them wonderful things about how comfortable Roman was with my car, and I guaranteed that they could take pictures from a distance of no more than five yards. However, Roman turned out to be shyer than usual on this day. This was not too surprising since now there were three men

---

**Homo sum; humani nil a me alienum puto* (I am a human being; I consider nothing belonging to human life as being unfamiliar to me).

leaning out of my car, and he was not used to this sight. He did not flee at a gallop, and he did not leave his territory, but he always withdrew to a distance of twenty to thirty yards. Since my "honor" was now at stake, I followed him slowly but persistently. Eventually, Roman stopped walking, bedded down, and allowed me to get within five yards of him. My guests took their pictures and expressed the opinion that Roman's behavior was magnificent proof of this buck's "personal trust" toward me.

I have to admit that Roman's lying down in this situation came as a surprise to me, and even today I am not quite sure how to interpret it. Perhaps the following may come close to the truth: to Roman, I or my car may have come to represent a somewhat unusual, certainly very big and strong, but otherwise harmless companion—some kind of a "super-gazelle." When I followed him so persistently that day and he obviously did not want to leave his territory, he may have exhibited a hint of submissive behavior. When a subordinate gazelle is threatened or bothered by a very dominant conspecific but for some reason is not able or willing to flee, the subordinate will sometimes lie down. In full submission, he stretches his head and neck forward flat on the ground. This Roman did not do, and thus, as I said, he showed only a slight intention of submission. Provided this interpretation is approximately correct, it may be said that Roman expected me to recognize and respect his submission, and this again may be assumed to be an indication of "personal trust," although in a somewhat different way than my guests had meant it.

Roman was in his territory from January through the beginning of July 1965. During the "long rains" (March through May), the grass had grown higher there. Since Thomson's gazelle prefer short-grass areas, the herds of females did not visit his territory as often as before. When the whole of Togoro began to dry up in June, they left. Quite a number of territorial males also left their territories at that time. The tommies still remaining in Togoro were predominantly adult bucks—some of them territorial, some nonterritorial. Now the owners of the territories displayed a striking tolerance toward the bachelors. For instance, a small herd of males was grazing inside Roman's territory. When one of them crossed or stood in Roman's path, he presented his horns toward him, and the bachelor immediately gave way. Thus, Roman clearly was still dominant over the nonterritorial bucks, but he no longer chased them out of his territory. Sometimes he even seemed to be interested in keeping their company. When they moved on, he peacefully followed them to the boundary, stopped there, and watched after them. Or he even accompanied them beyond the boundary, turning around after a while and returning alone. On one of the first days of July, he left.

But Short-tail remained. He had shown the same tolerant behavior toward bachelors as Roman, and now that all his territorial neighbors had left, he appeared to be no longer sure about the position of the boundaries of his territory. Quite often he could be seen somewhat outside it; in particular, he frequently ventured into Roman's former territory, eventually using half of it as if it were part of his own. His marking activity dwindled drastically. In the absence of neighbors, he did not have any fights. When—very infrequently—a few females entered his territory, he herded them "dutifully," but he usually did not drive one of them up to urination and Flehmen, not to speak of more intense courtship.

During the rainy season, Thomson's gazelle do not drink water. They take in sufficient fluids from the plants they eat. However, they commonly drink during dry season. Though there was no water in Short-tail's territory and its vicinity, he did not leave it to drink. Throughout the whole observation time, I met Short-tail only twice outside his (now enlarged) territory or its immediate vicinity. Once he marched more than a mile to a place which at other seasons the gazelles frequented to eat (moist) soil. At that time, however, neither water nor any other moisture was available there.

The other time, as far as I could figure, Short-tail had had no contact with any conspecific for one full week, and during the week before only very seldom. One day a big mixed tommy herd, males and females, migrated through Togoro. The path of the herd took them more than half a mile east of Short-tail's territory. At first he stood highly erect at its boundary, watching the herd intently. Then he suddenly lowered his head to the right and to the left, lifted all four feet alternately, tossed his snout toward his flank, scratched his neck, and shook his flank—in short, he showed the behavior typical of the inner conflict a subordinate male may show when he is slowly approached by a dominant buck and is unsure whether to stand still or run. Finally, Short-tail ran at a gallop toward the herd. On his way, he stopped several times, each time repeating the whole sequence of conflict behavior. He arrived at the herd and intermingled with the other gazelles without any complications.

I followed him by car to watch at close range what was going on there. I could not detect any difference in Short-tail's behavior as compared to that of the other males, and I presumed that he would leave the area with this herd. As I had observed several times, when during dry season the majority of the gazelles have been gone from a given area for quite some time, but one day a migratory herd passes through it, this herd may "collect" the few remainders and take them along with it.

For about two hours, the herd, with Short-tail right among the others, remained grazing more or less at this spot. Then they moved faster ahead. When the move began, Short-tail fell behind. As abruptly as before, he performed the aforementioned sequence of conflict behavior, turned around, and returned to his territory at a gallop. Apparently the attraction of the conspecifics and their company had been in conflict with the tendency to stay in his territory, and the latter urge was still preponderant.

Also on other occasions, when a buck has been solitary in his territory for quite a while, his dammed-up need for interaction with conspecifics may lead to somewhat unusual scenes. For instance, once a small herd of male Grant's gazelle entered Short-tail's territory. Generally, there is nothing unusual about that. The territorial intolerance of tommy bucks—as well as that of territorial males in all the African ungulates known to me—is almost exclusively exhibited toward conspecifics of the same sex. (In zoological gardens, it can be different.) Animals of another species, including the males, are tolerated within a territory. The territories of males from different species may even overlap completely or in part. For example, a tommy buck, a male Grant's gazelle, and a topi bull may have their territories at the same place. When none of them is visited by females of his own species for some time, the three males often graze, move, and rest together.

This latter occurrence is somewhat striking. The so-called "socialization" of African ungulates of different species usually goes only so far that they do not avoid each other. Therefore, they may occasionally come together at a water hole, in the shade of a single tree, in a locality with particularly good grazing conditions, etc. This has as little to do with true socialization as the gathering of people at a railway station. The other guy is neither avoided, nor feared, nor attacked, but otherwise he is little more than some kind of obstacle.

In Grant's and Thomson's gazelles, there is a somewhat stronger reciprocal attraction. These two species act like they like each other, and thus they may frequently be seen grazing, moving, and resting together. Other social interactions, however, are rare also among these two species. As stated above, there is no territorial competition between them. Also, I have never seen a tommy buck courting a Grant's female, or vice versa. Only short aggressive interactions among members of the two species—for instance, over a particularly favored food at a very limited place—may occur on rare occasions in the wild.

Thus, as I said, there was nothing special about it when a few Grant's bucks entered Short-tail's territory and grazed near him for a while. However, when they moved ahead and were about to leave his territory, strangely enough, Short-tail tried to stop them and herd them back from the boundary toward the center—as if they were tommy females. It hardly needs mentioning that the Grant's bucks were not very cooperative. Zoo animals with no or only a very few conspecific partners available to them have been known to exhibit certain behaviors normally performed only with conspecifics toward animals of another species (including humans). My admired and beloved teacher, Dr. Heini Hediger, had spoken of an "assimilation tendency" in such cases, and he considered it to be caused by the captive conditions. Short-tail demonstrated quite impressively that this assimilation tendency is not necessarily restricted to animals in captivity but may also, though considerably more rarely, occur in the wild.

Short-tail kept his territory up until the middle of December, for almost a full year. This was the longest uninterrupted stay in a territory which I ever recorded in a tommy male. Three months before he left, a number of gazelles returned to Togoro. Upon their arrival, the "mild" territorial activity which Short-tail had shown during the dry season before became abruptly pronounced. He again began to chase all non-territorial males out of his territory, he herded and courted females, he had boundary encounters with territorial neighbors, and he marked frequently by means of preorbital gland secretion and by urinating and defecating on dung piles. In short, he behaved as a "good" territorial tommy buck is expected to do. In mid-December, suddenly he was at the end. Although the whole of Togoro was covered with fresh, green short grass and gazelles were there in abundance, one day he left his territory and the entire area. Perhaps his "territorial energy" had been largely consumed during the long stay in his territory through the dry season and finally had become exhausted after the return of the herds.

Ten months later, in October 1966, Short-tail was seen again in Togoro. He arrived there with a big mixed herd, which left the area on the same day since there had been no rains yet in Togoro at that time of the year, and the land was still bone-dry. Short-tail, however, stayed behind, along with two other adult males, remaining pretty precisely inside his former territory. He began to mark there intensely, and eventually he chased away his two companions. He "played territorial buck" for three days, then left the area for good.

With the last descriptions, however, I have gotten ahead of the story. One year before, in September 1965—Short-tail was still present and occupying his own and half of Roman's former territory—there were some extraordinary rainfalls in Togoro (extraordinary insofar as it still should have been dry season according to the calendar). As always after a rain in the African plains, short green grass started growing immediately, and, attracted by it, the number of gazelles increased in this area.

Among the newcomers, there was a tommy buck I named "Kimm." He set up a territory next to Short-tail's, including the other, still unoccupied half of Roman's former territory in his territory. Where the boundary between his and Short-tail's territories was now located were the remnants of a broken tree trunk. Back in Roman's day, a "neutral edge" had developed around this spot. When bachelors retreated to this place, Roman would let them stand there in peace. Short-tail and Kimm also treated the area immediately surrounding this broken tree as a "no-man's-land."

During the second half of October 1965, the "small rains" began with full force. The whole of Togoro was a green meadow, and the gazelles arrived there in large numbers. Meanwhile, I had extended my observations to other parts of the Serengeti National Park; however, I still visited Togoro at least once a week, since I did not wish to lose track of things in this area so familiar to me. Admittedly, I did not feel as much pleasure in going there as in former times. I missed something there since Roman had left.

So I came to Togoro again on one day at the end of October 1965. I left "the road" and passed "the little forest," as I had named a small group of acacia trees in the otherwise open plains, on my way to Short-tail's territory. A single tommy buck rested among the trees of "the little forest." "Perhaps a new territorial male, perhaps a single bachelor. Well, I can look after him later. First, let's go to Short-tail." In the middle of my monologue, I stepped so heavily on the brakes that, despite my low speed, all of my things went flying around in the car. It was Roman!!

With great excitement, I dug in my old notes until I found Roman's "warrant." With binoculars at my eyes, I compared the features of this buck to the data I had noted down nine or ten months ago. The black spot on the nose, the white spot on the forehead, the dark stripe running from the eye to the angle of the mouth, the length and the shape of the horns, the relative length of the tail, etc., everything matched with my notes. For a final "test," I turned my car and slowly approached the resting buck—up to thirty yards, twenty yards, ten yards, five yards. He arose, stretched his body and his hindlegs, yawned, scratched his ear, shook his head, and then began to graze peacefully not quite five yards in front of me. Only one tommy buck felt so at ease with my car. No doubt, this was Old Roman, who obviously had returned to Togoro with the migratory herds, had left them, and was now only about three hundred yards away from his former territory, which meanwhile had been divided up between Short-tail and Kimm.

In my thoughts, I canceled everything I had intended to do in the near future. Originally I had planned to stay in Togoro only until evening. However, as was my routine, I had sufficient water and food for three days with me. If I used these supplies with some care, I could make them last even longer. I did not want to miss the events which were to be expected now.

After Roman had grazed awhile between the trees of "the little forest," he marched briskly and confidently toward his former territory. He entered the half now occupied by Short-tail. Immediately, Short-tail approached him at a gallop with loud "chasing calls." Roman fled from him into the half of his former territory now occupied by Kimm. Promptly he was chased by Kimm, and, in fleeing, he came back to Short-tail's area. The latter chased him across his entire territory. Finally, Roman fled back to "the little forest," where he was not pursued and could recover his breath. Later on, he tried several times to return to his former territory, but always with the same negative and to him certainly most unpleasant result. The next day brought only a small change, in that Roman did not flee to "the little forest" anymore but retreated to the "no-man's-land" around the fallen tree, where Short-tail and Kimm tolerated his stay.

By the third day, Roman had established a kind of "mini-territory" there. The beginning of a dung pile was recognizable, and several times he chased away bachelors who had tried to escape from Short-tail or Kimm into this formerly "neutral edge." In the evening, he pursued a fleeing bachelor far into Short-tail's territory. Immediately, Short-tail charged him at a gallop—and Roman stood and fought back. The fight was vehement but absolutely according to the rules of encounters between territorial neighbors, ending with a grazing ritual. On the next day, several encounters of this type took place between these two bucks, and on the fifth day, Short-tail retreated to the former boundary of his original territory without being defeated by Roman in any way. Thus he gave up the area he had occupied during Roman's absence, and Roman had won back half of his old territory.

For a few days, Roman appeared to be content with this situation, and he had only the usual boundary encounters with Short-tail and two other bucks who had established their territories to the right and to the left of his. But then Roman tried to reconquer the half of his former

"Roman" and "Kimm" fought almost constantly.

territory now occupied by Kimm. Here, however, he met a bitter resistance. I do not consider it too anthropomorphic to say that each of the two bucks felt himself to be "in the right"—Roman because this area had formerly been part of his territory, and Kimm because he had occupied it at a time when no owner was present.

All the encounters between Roman and Kimm took the customary course and were ended by grazing rituals. However, their fights were always long-lasting, often very vehement, and above all extraordinarily frequent. On one day I counted forty-three fights between Roman and Kimm between sunup and sundown. Initially I had in mind to determine the total number of their encounters; however, I soon dropped this idea because, speaking with only slight exaggeration, these two were permanently occupied with each other. At night, before falling asleep in the sleeping bag in my car, I could still hear the clashing of their horns, and I awoke the next morning to the same "music."

At least on Roman's part, these encounters were not the usual "boundary ratifications" between territorial neighbors, for he tried to gain space by carrying the fights ahead from one small landmark to the next. At first he was fighting over a small termite mound situated about ten yards beyond the boundary in Kimm's territory. Having conquered this section after a couple of days, he then tried to push forward toward a little isle of high grass about ten yards away from the termite mound. After he had won some space in this way, however, he lost all the conquered terrain to Kimm again, whereupon he started over from the beginning.

With advances and retreats, Roman's efforts to reconquer his former territory lasted from November 1965 through April 1966. Once he was wounded in a fight. Of the one thousand fights between gazelles that I saw and recorded in Serengeti, this was the only one involving bloodshed. It was an accident. In vehement fighting, Kimm delivered a forward-thrust attack, which an opponent normally parries with his horns. This time, however, the horns of the adversaries glided off each other, and in the swing of the attack Kimm pushed forward below Roman's head and neck, tearing a long wound in the latter's chest with his horns. Roman brought the fighting to an end and retreated into the portion of his territory that he had regained from Short-tail, where Kimm never followed him. During the subsequent days, Roman avoided all encounters with his neighbors. He even fled from Kimm a few times when he met him near the boundary. After two weeks, he had recovered and started fighting Kimm again.

One day in the beginning of April, Kimm disappeared. I do not know what happened to him. Roman occupied that part of Kimm's territory which had formerly belonged to him. The other, about equally large, part of Kimm's territory remained empty for weeks until another tommy buck established a territory there.

A short time later, I was parked once again in the middle of Roman's territory. It was a peaceful evening. Hundreds of gazelles were grazing around my car, Roman close by as usual. Since it was the time of the full moon, I had prepared myself for a night observation. When the sun was already down to the horizon, suddenly a pack of wild dogs showed up on top of the next (low) hill, running at full speed, and immediately started to hunt gazelles. The gazelles obviously were as surprised as I was. They dashed to and fro in confusion as if a bomb had exploded among them. A wild dog on his heels, Roman jumped up in an enormous stotting gait*—the highest stotting I have ever seen in a gazelle—

*For information on stotting, see chapter 7.

flying over the hood of my car. The dog had to run around the car and lost Roman in the crowd.

I had the general impression that these wild dogs were somewhat confused by the large number of gazelles, and perhaps also by their stotting. They did not concentrate their hunting actions on one particular prey animal, as I have observed on other occasions, but each of them hunted individually for a gazelle. When another gazelle crossed one's path or came otherwise closer, the dog let the first selected prey run and began to chase the other. Finally, all the gazelles had disappeared from the area, and only the wild dogs stood there, with their tongues hanging out of their mouths and obviously out of breath. They had not caught anything. They lay down and rested for a while. Finally they arose, lined up in single file, and trotted away.

Meanwhile it was night—as expected, with a splendid full moon. In the far surroundings, all the game animals had gone, and the chances for some good nocturnal observations were gone with them. I ate something, smoked a pipe, played the harmonica, and moved some more or less wise thoughts through my mind. At about ten P.M., three hours after the attack of the wild dogs, I recognized movements here and there in the dark. The gazelles were returning to the area; of course, the territorial bucks came first of all, each heading for his territory.

Soon Roman returned, too. He entered his territory, approached my car, and lay down right beside the driver's cab. When observing animals for a period of hours, I always dismantled the upper half of the car's door (this is possible with a Land Rover). This time, however, during the long pause in observation, I had opened the door completely and had stretched my legs out of the car. Thus, I was sitting there, fully exposed, and not quite two yards away a free-ranging gazelle buck had come and bedded down completely on his own and was now looking out into the moonlit night with his back toward me. I felt this hour to be one of the solemn moments of my life.

In the second half of May, Roman showed the characteristic behavior of a tommy male whose territorial period is coming to an end. One morning in June, his territory was empty. I had still seen him there the evening before. Thus, he probably had not yet gone far away. I had a few matters to take care of that day which I hardly could postpone, but I started searching for him at noon. I discovered him about two miles away from his territory in a bachelor herd which, though somewhat changed in terms of composition, had been in the area for some time. By the next morning, this whole herd had disappeared. Again I drove all over Togoro in a systematic search. Finally, in the woodland surrounding the plains, near a korongo (a creek which contains water only during rainy season) and about four miles away from the place of his last stay, I found Roman in a big mixed tommy herd, a typical migratory herd.

The animals were grazing. I slowly and carefully maneuvered my car between the trees and bushes toward this herd. When I got within thirty or at most twenty yards, all the gazelles withdrew from my car; only Roman allowed me to approach him, to within five yards. At ease as always, he grazed beside my car for a while. Then he followed the herd, which was now moving resolutely along the korongo deeper into the woodland. His territorial time in Togoro was over. The long move had begun.

Near Roman, I followed the gazelles over some five miles. Then the country became extremely rough, even by African standards. Since I badly needed an intact car for my research, I did not want to take a risk. Anyway, I could not follow the animals for days, weeks, and months. Thus, I eventually stopped.

Between trees and bushes glimmering in the sunlight, Roman disappeared from my sight—the tommy buck to whom I owed so much insight into animal life and such wonderful personal experiences, and to whom I, too, had apparently been something more than an inevitable evil.

I did not see him again.

AT RED WATERS

Ngare nanyuki, which means "red water," is the name of a place in the Masai language. For some reason, African waters may sometimes appear reddish in color seasonally and/or locally. On modern maps of the Serengeti National Park, there is a whole river with this name. In the old days (1965/66), "Ngare nanyuki" meant only the spot where a "road" crossed a korongo (a creek without water during dry season), a tributary of said river. As is common in Africa, the name did not refer only to the exact point itself, it also included the surroundings within a radius of several miles, so that nobody knew precisely where the named area began and where it ended. In my private geography, "Ngare nanyuki" meant all the land from the creek with its gallery forest over about five miles of completely open plains, ascending in erosion steps toward the Barafu kopjes, an impressive group of rocks.

To avoid any possible error: There is no human settlement. Ngare nanyuki is simply the name of an area near the eastern boundary of the national park, about thirty miles (as the crow flies) from the headquarters in Seronera. To get there, one must take the way from Seronera

High noon at Ngare nanyuki during dry season.

toward Loliondo, which was characterized as a motor road on the maps in 1965. This was a charming exaggeration, since this "road" consisted only of a car track that was not always well visible. Correspondingly low was the "density of traffic." Altogether, I spent more than one hundred days at Ngare nanyuki, and did not see other people more than five times.*

Certainly, life was pleasantly undisturbed out there. On the other hand, when a man's car broke down and he could not repair the damage himself, he was in trouble. I once had this rotten luck. The next human settlement was Seronera. A walk of some thirty miles is not an extraordinary performance, but neither is it a pure pleasure, particularly when the land is populated not only by graceful gazelles but also by rhinos,

---

*Recently I was told that nowadays (1993) even this formerly so valuably lonesome area is no longer safe from mass tourism.

lions, and hyenas. Especially when thinking of the hyenas, I felt a little uncomfortable. Spotted hyenas are numerous in Serengeti, and they sometimes can become pretty importunate. That they exclusively eat carcasses is a pious legend. They kill everything they can get, up to the size of a fully grown zebra or wildebeest.

I decided that I would walk only when there was nothing else left. For the time being, I was quite well off. I had enough food and water for several days, and there was a lot to observe with the many game animals around me. So the time I spent there would by no means be wasted. I also was only about two miles away from the "road." Anyone who passed by would have to become aware of me, and I would see him.

I was not so bold as to hope that a car would just happen to come along. However, Nyafi Salewa, my cook and "jack-of-all-trades," was waiting for me at my house at the research station Banagi. Officially he called himself a cook, but in this regard he was not much better than I myself—and my culinary arts are not overwhelming. As a "jack-of-all-trades," however, he was incomparable. I had told him where I intended to go and how long I planned to stay there. So I hoped that he would prove himself an intelligent person in this regard as well and eventually would sound the alarm.

He did so after three days and went to Dr. Cain from Cambridge University (in England), who was the only scientist besides me at Banagi at that time, and also the only man there who had a car available. Arthur Cain, however, answered that there are hundreds of reasons why a field worker may not return on time, and that it therefore might be best to wait a few more days. At first, Nyafi made a long pause of "hard thinking"—as he always used to do when somebody had said something he did not agree with—but then he declared emphatically: "The bwana swala [mister gazelle—a surname given to me by the people in Banagi because I was working with gazelles, and because they had problems pronouncing my name anyway] is always on time. When he has said 'I'll be back after three days,' and he is not back, he must be in big trouble." This was enough to convince Arthur Cain. Next morning, he started the search for me. He greeted me with the words: "Dr. Walther, I presume!"

The area near the korongo of Ngare nanyuki, with its fringe of forest containing several (natural) clearances, was really beautiful. Once I was visited by Morris Gosling, a young British scientist, who later conducted an excellent study on the behavior of kongoni (East African hartebeest) in the Nairobi National Park. I took him with me to Ngare nanyuki, and when we arrived there, he exclaimed: "This is the Africa I have dreamed of since I was a boy!" Indeed, it was "Africa out of a picture book."

Except at the height of the drought, kongoni, topi, and warthog were always found in the gallery forest, and frequently a few rhinos as well.*

One or two miles after crossing the Ngare nanyuki creek, I usually turned off "the road" and headed south. Soon I had left the forest behind me, and now the seemingly endless short-grass plains began, expanding to the Ngorongoro Crater far away at the horizon. This land is completely open but not flat. It is "rolling country," rising and sinking in low "waves," and very dry most of the time; the several river beds that dissect it contain water only during rainy season. Trees do not grow along these korongo in the open plains, but a few bushes may be found scattered along their fringes. Typical of the landscape are also a few "kopjes," groups of rocks which tower like islands above the plains.

During the rainy season, thousands of wildebeest, zebra, and Thomson's gazelle populate the country. Herds of eland antelopes and smaller groups of some other game species are also common. Except for ostriches and a very few single tommy bucks, all of these animals leave during the long dry periods. Instead, now oryx antelope may occasionally show up. All year round, however, this land is a stronghold of Grant's gazelle. At the high point of the drought, some of them may leave, but they never disappear from this area entirely.

It was because of the Grant's gazelle that I came so often to Ngare nanyuki. The largest herds and the strongest bucks in all of Serengeti were found there. That this was not just my private opinion was confirmed by one of the very few occasions when I met another man out there. One day, a Land Rover appeared on "the road" from the park boundary, heading in the direction of Seronera. It stopped, and the driver left the car and looked around with binoculars. I was parked only a few miles away from "the road," and I drove closer to catch a glimpse of this guy. Upon arriving, I recognized an old friend. Ulrich Trappe had been born in East Africa about fifty years before, the son of German parents. He had spent most of his life as a hunting guide and a game warden, but at that time he was with the veterinary research station at Kirawira in "the corridor" (the western extension of the Serengeti National Park toward Lake Victoria).

We greeted each other. "Where are you coming from?" I asked him. "From Nairobi. I was shopping there." "And you returned via Loliondo?" "Yes." I looked at him out of the corner of my eye: "Well—this is not the

---

*Twenty years later, almost all the rhinos had been poached in the Serengeti National Park.

shortest way to Kirawira, and by no means the smoothest!" He smiled, a little embarrassed: "I think I can tell you why I did it. You will understand. I took this way because I once spotted a Grant's buck out here the like of which I have not seen three of during all my many years in East Africa. With some luck, perhaps I may see him once more." I knew of whom he was speaking. This was "my" Grant's buck, "Wide Horn of Ngare nanyuki."

The Grant's gazelle in Serengeti belong to a subspecies in which the horns of the fully grown males curve to the outside with the tips pointing somewhat toward the ground. This is the Wide-horned Grant's gazelle or Robert's gazelle. They are considerably taller than Thomson's gazelle, about the size of a white-tailed deer. In modern Kiswahili, they are called "Swala granti." This is the literal translation of the English "Grant's gazelle." I wondered whether this species had not had a special name before the explorer James August Grant traveled in the country about 1865. I could not get any information from my African friends and acquaintances. Finally, in an old book published before World War I, when Tanzania was still a German colony, I read that the natives had used the name "swala nyeupe" (pronounced "nyai-oope") at that time. Obviously, this name has disappeared completely from the vocabulary of modern East Africans. "Swala nyeupe" means "bright gazelle," or even "white gazelle." (The meaning of adjectives is often broad in Kiswahili, and "nyeupe" may be used for any bright color, including white.)

If this is correct, the name was well chosen. When I was making color sketches of African antelopes in the wild, it often struck me that there are several species—and Grant's gazelle definitely belongs among them—whose coat color changes considerably with the direction of the light and the color of the background. Depending on both, Grant's gazelle may appear to be orange-colored over all shades of light brown up to almost white. Most commonly, the color of their coats corresponds to a pale ocher, resembling a palomino horse. Thus, they truly are "bright gazelles."

The behavior of Grant's gazelle is similar to that of Thomson's gazelle in general, but often different in the details. For instance, territorial Grant's bucks establish dung piles, too, although not as many as tommy bucks in their much smaller territories, and all Grant's males perform object aggressions and scrape the ground, urinate, and defecate in the same sequence as tommy bucks, with the same movements and in the same striking postures. In contrast to the latter, however, they do not mark grass stems, small branches, or any other environmental objects with the secretions from their preorbital glands, even though

The Grant's gazelle in Serengeti belong to the subspecies of the wide-horned Grant's gazelle (*Gazella granti robertsi*).

they possess such glands and, as far as I could figure, these are functional.

Generally, the activities of Grant's gazelle often appear to be more sedate, more "dignified," than those of the much smaller and always busy tommies. When Grant's gazelle move over the plains, it sometimes looks almost as if they were moving "on rails." Of course, they are able to turn on the spot like any tommy, but they prefer to walk in a small semicircle when changing direction. Otherwise, they show the same gaits as Thomson's gazelle.

Also, the mating rituals of these two species are like "two variations

The foreleg kick is reduced to a big, stiff-legged step in the courting Grant's buck.

on the same theme." As in Thomson's gazelle, the courting Grant's buck follows the female at a walk. However, in doing so, he stretches his tail horizontally backward, which a tommy buck does not do. In the first phase of courting, the Thomson's male approaches the female with his neck and head stretched forward. Only very occasionally can a similar posture be seen in a male Grant's gazelle in the same situation, but here it is anything but obligatory. It is doubtful whether this is even a specific display in this species. Grant's and tommy males both perform foreleg kicks in courtship; however, the Grant's buck's kick is reduced to a big, emphatic step with a stiffly stretched foreleg. This move may also be seen in the tommy, but in addition to at least two other forms (lifting the foreleg up to almost horizontal level and alternating kicks with both forelegs), whereas it is the only form in Grant's gazelle. During the entire

Mating march in Grant's gazelle.

mating ritual, a Grant's male keeps his neck erect and his nose lifted above horizontal. In the tommy, only a momentary movement, a "nose-lift" or "nose-up," of the male at the beginning of the ritual corresponds to this attitude continually maintained by a courting Grant's buck.

When driven by a male, a Grant's female behaves essentially the same way as a tommy female; however, she is more aggressive. Occasionally, she may even turn around and briefly fight the buck, whereas I have never seen a female Thomson's gazelle fight an adult male. After being driven for a while, a Grant's female often throws her head up and sideward-backward in a jerky fashion. This resembles a movement in the dominance display of Grant's gazelle (to be described later) and is probably a somewhat modified form of it. The Grant's buck usually reacts to it by mounting immediately. In Thomson's gazelle, there is no parallel to this display of the Grant's female.

At the beginning of a mating ritual, a "high-spirited" Grant's female shows further dominance displays. I suspect that these particular females may be predominantly young adults, courted by a male for the first time in their lives. To the human observer, the mating ritual of Grant's gazelle always means an aesthetic delight. With such a "high-

After being driven by the male for some time, a Grant's female may throw her head up and sideward-backward in a fashion resembling head-flagging in hostile encounters. The buck usually responds to the female's display by mounting.

spirited" female, it becomes a true ballet. With his head lifted, the buck approaches the female from the front with an exaggerated, stiff-legged step, a ritualized foreleg kick. In response, the female first retreats by stepping backward, assuming an erect posture. In erect displays, the male and the female move toward each other and arrive at a reverse-parallel (head-to-tail) position in which they circle around each other. Occasionally, the female may now take a frontal position toward the male and fight him. After a short fight, horn to horn, the female breaks away with a jump. These "figures" may be repeated several times in the same or another sequence. Finally, even the most "high-spirited" female stops her counterdisplays and walks ahead, followed by the buck in erect posture.

Like a tommy buck, a Grant's male may herd the females in his territory and try to prevent them from leaving it. The territories of Grant's gazelle, however, are considerably larger than those of tommies, often ranging between fifty and one hundred fifty acres. In contrast to

At the beginning of a courtship ritual, a Grant's buck and a female may move toward each other in erect displays, finally arriving in a reverse-parallel position in which they circle around each other.

Thomson's gazelle, on a mbuga* in the woodland, a stable "harem" group of commonly five to twenty Grant's females will stay with one buck in his large territory for weeks and months. Only the buck is territorial in these cases, however, not the females, and only he is aware of the boundaries of his territory. The females would cross them without hesitation if not stopped by the male. If they deliberately try to leave him, he will attempt to block their path and/or to chase them back toward the center of the territory. These actions are as striking as those of a herding tommy buck. However, it took me a shamefully long time to become aware that within his territory, a Grant's buck almost always signals to his females where to go or not to go. He does it in moving, standing, grazing, and sometimes even in resting, and he accomplishes it merely by his position and orientation with respect to the females. The

*The basic meaning of the Swahili word *mbuga* is "steppe, grassland, open plains." Sometimes, however, the word is used in a more special sense meaning a small or large (natural) clearance in a forest, an open plains area surrounded by woodland. The word is used in this sense in this book.

In the beginning of his courtship, a Grant's buck sometimes frontally approaches the female, who may respond with an erect counterdisplay.

only presupposition is that he is not right among them but a few yards separated from them, as is frequently the case. He usually directs his actions only toward one female, the one farthest ahead in the desired or not-desired direction.

The easiest way to describe this signaling system may be to "translate" the meaning of the buck's positions into words of the human language. For instance, when the buck positions himself broadside in front of the female (lateral T-position of the male), this means "Stop!" When he stands broadside behind her or reverse-parallel beside her, this means "Go where you like except in the direction blocked by me!" When the buck stands or moves parallel beside the female, he signals "Keep your intended direction and walk somewhat faster!" Even stronger is his demand that the female speed up while maintaining the same direction, which the buck signals by orienting himself frontally toward the female's flank (sagittal T-position of the male). When the buck frontally

HERDING BEHAVIOR IN GRANT'S GAZELLE

Courted by the buck ("Wide-Horn of Ngare nanyuki"), a Grant's female is approaching the boundary of the male's territory.

The buck passes her, . . .

assumes a broadside position in front of her, blocking her path, . . .

. . . and herds her back.

approaches a female in head-on position, this means "Turn around immediately and move on!" When he stands or moves in tandem position behind the female, it means "Go ahead!" (usually without a special directional request). And when he positions himself in tandem position in front of her, it means "Follow me!"

These are a few "vocables" from the "dictionary" of Grant's gazelle. They can be modified by combinations (for instance, when the male stands obliquely in front of the female, a position in which frontal and broadside orientations are mixed), and they can be completed and reinforced by expressive gestures and postures (predominantly threat, dominance, and courtship displays). As I said above, however, often the buck's positions and orientations are sufficient in themselves to direct the course of the females moving inside the large territory. This "silent herding" is inconspicuous but it works very effectively.

In Grant's as well as Thomson's gazelle, the young lie out* (they are separated from their mothers at rest most of the time) during the first days and weeks of their lives, and there are no striking differences in the mother-offspring relationships of these two species. Perhaps "mothers' groups," groups of females all having fawns of about the same young age, may be somewhat more frequent in Grant's gazelle (although still not very common), but they do exist in the tommy as well.

In general, the social organization is very much alike in both species. In Thomson's as in Grant's gazelle, there are mixed herds of males and females of all age classes except fawns less than half-grown. In addition, the bucks often form all-male groups ("bachelor" herds). Youngsters from adolescence (about seven to twelve months in Grant's gazelle, about five to eight months in Thomson's gazelle) up to males of a very old age (probably beginning around nine to twelve years) can be found in them, but usually bucks in their prime make up the majority of individuals in the all-male groups of gazelles. Correspondingly, females form all-female groups with females of all ages, with or without offspring—the "mothers' groups" mentioned above are a special case. (Because female groups without fawns are by no means rare in the two East African gazelle species, I do not think that the term "nursery herds"—frequently used by other authors—is appropriate.) Adolescent males may still be found in female groups, but they also may have already switched to all-male groups. In any case, there are no males beyond the age of adolescence in a female group. Shortly before a female gives

---

*For more information on lying out, see chapter 6.

birth, and often for a few days after, she separates from the others, usually simply by no longer participating in the movements of the herd, and thus remaining behind when the group moves ahead.

As in Thomson's gazelle, some adult Grant's males leave the herds and become territorial, but their territories are about ten times larger than an average tommy territory, and they are often separated by strips of "no-man's-land." Therefore, boundary encounters and fights among territorial neighbors are much rarer in Grant's than in Thomson's gazelle. As in the latter, groups of females may sometimes visit the owners of territories for only a few hours a day. More commonly, however, in Grant's gazelle on a mbuga in the woodland, a group of females stays in one territory with one male throughout his entire territorial period, which easily may range between five and eight months at a time.

An all-male group containing numerous adult bucks with their long horns is a marvelous sight. Such a bachelor herd moves through its very large home range in a daily "circuit." At Ngare nanyuki, for several weeks, a herd of more than sixty bucks regularly came to one particular place in the late afternoon hours. This place was only about half an acre in size, and it differed from the surroundings only insofar as the grass was especially scarce there and the ground was largely covered by sand. I termed it "the lists." Upon arriving there, the bucks began to fight. Sometimes as many as twenty pairs were fighting simultaneously. One pair started, a second pair "became infected," and then a buck not previously involved "took offense" and intervened by attacking one of the two combatants. The weaker male broke away from this fight, immediately rushing into another buck with whom he continued fighting. At first the "victor" followed him, but soon he, too, became involved in a struggle with another male. Meanwhile, two pairs started fighting at the opposite end of the herd—and so on. This spectacle lasted from ten minutes to half an hour per day, and very probably there was not one among the numerous bucks who had not fought at least once during this time.

Although I was well trained in speaking my observations into the tape recorder and taking photos of the events at the same time, both with quite remarkable speed, I often did not manage to keep track. It was one of the very rare times I regretted not having an assistant with me. For one thing, I never found the time to change the lenses of my camera, and since I customarily had a big telephoto lens on it, I was able to take some good pictures of the single fighting pairs but none showing several or all of them. Nowadays I feel very sorry about this, and I can only hope that the reader will trust my words without documentation by photos.

BEHAVIOR AT "THE LISTS"

Above: Upon arriving at "the lists," the Grant's bucks begin to fight.

Below: A third buck "takes offense" and intervenes in the fight between his two companions.

Of course, these fights offered an excellent opportunity to study the fighting behavior of Grant's gazelle. These animals use a number of tactics or techniques. Some fights are carried out according to only one of them, while in other encounters, several or even all of them may be employed. Some of these tactics typically occur at the beginning of a fight, some of them predominantly toward its end. Furthermore, there are certain defensive maneuvers which are precisely adapted to particular attacking movements. Thus, it is possible to report on such a fight much as a radio reporter speaks of a fencing or a boxing match between humans.

One fighting technique, often used at the beginning of an

encounter, is horn-pressing. The opponents stand or approach each other frontally with heads lowered in an attitude similar to grazing. Quite frequently, they first touch each other with their noses. Then they lean the full length of their foreheads and horns against each other, and each of them tries to press the rival's horns back toward the latter's nape.

In every fight of any length, the horn-pressing changes into forehead-pressing, which is the most common and important fighting technique in Grant's gazelle as well as in many other horned ungulates. The forehead of a fighter is lowered, often close to the ground, and sometimes even right on it. His chin is pulled toward his throat so that his horns point more or less forward, in a pronounced case horizontally. The rivals cross and interlock their horns near the bases, their foreheads almost or literally touching. The head of each of them may be between the other's horns like a horse between the shafts of a cart. Firmly "anchored" in this way, the fighters press against each other with all their might. Thus, the horns are used not so much as weapons but more as clasps which lock and hold the opponents together.

Yet another tactic is twist-fighting. With horns firmly interlocked, the bucks twist their heads alternately to the right and the left, each trying to superimpose his rhythm on that of his adversary. Apparently, this is a quite "convincing" demonstration of superior strength. However, in the many fights I observed, it never happened that a fighter sprained or broke the other's neck in this way—which certain theorists have presumed to be the purpose of this action.

When their horns are anchored in a forehead press, one of the fighters may push forward heavily with his hindlegs close together, his forelegs widely spread, and his back arched. He puts his whole weight behind the push and, so to speak, "rolls" his body's center of gravity over his shoulders and nape forward-downward into the clash. Occasionally both rivals will press against each other in this way. More frequently, however, it is only one of them who attacks in this form, and the defender tries to parry the attack by ducking deep down. He stretches his neck and head forward-downward so that his horns lean back toward his nape, springing so far down with his widely spread forelegs that his chest almost touches the ground, a maneuver which works well to catch the other's attack and let it run out. When this is effective, the defender is in a good position for a powerful counterattack. If a buck is stronger than his opponent or at least as strong, he sometimes may assume the deep-ducking posture before horn contact is established. He invites the other to attack so that he can have the oppor-

FIGHTING TECHNIQUES OF GRANT'S GAZELLE

Above: Horn-pressing. *Below:* Forehead-pressing with horns interlocked.

tunity for an even stronger counterattack. Of course, such a counterattack is by no means according to the first attacker's intentions. Therefore, he may try to overrun the opponent's deep-ducking defense by delivering a forward thrust. Immediately after the attack with arched

Forward-pushing (*left*) parried by ducking deep (*right*) in horn-pressing.

back, he stretches his back to the maximum and steps forward with his hindlegs. Now the defender must jump backward as fast as he can in order to regain a firm stand. If he is not fast enough, he will be pushed back over quite a distance so quickly and vehemently that he will inevitably stumble, and then the only thing left to do is to break away and flee at a gallop.

In forehead-pressing as in twist-fighting, it sometimes happens that the opponents lock their horns so tightly that it is difficult to break loose. In this situation, they pivot wildly with their hindquarters while keeping their heads in the circle's center, twisting their heads, and pushing and pulling with their horns vehemently. As soon as they get free from each other, one of them turns around and flees. Thus, this fight-circling usually occurs at the end of a fight—if it occurs at all. The human observer may get the impression that it is a rather desperate maneuver. However, I have never witnessed and never heard that Grant's bucks were not able to break loose eventually, and that the rivals remained locked together so that both had to die—as may happen occasionally, for instance, in kudu. Thus, I consider it unlikely in Grant's gazelle. On the other hand, I would not be surprised if horns were sometimes broken as a result of fight-circling in this species, although I have to admit that I never observed a concrete case of this kind either.

In principle, the fighting techniques of Grant's gazelle are no different from those in Thomson's gazelle, but there are differences as to frequency. For instance, forehead-pressing and twist-fighting are more frequently seen in Grant's than in tommy bucks, whereas clash-fighting (in which the fighters flatly jump toward each other with all four legs, cross their horns in a short but violent clash, then leap back and immediately forward again for another clash—altogether, a modified form of forehead-pressing) is the most common type of fighting among territorial tommy bucks but is much rarer in Grant's gazelle.

The threat displays of the two species also show similarities. Both Grant's and Thomson's gazelle display a medial (head on body level) and high (head above body level) presentation of horns, and the head-low posture (similar to the grazing posture). The latter is more frequently seen in Grant's than in Thomson's gazelle, whereas the high presentation of horns is considerably more frequent and often better pronounced in tommy bucks. The opposite is true for sideward angling of horns. This is relatively rare in the tommy but quite typical of Grant's gazelle. The buck stands with his flank toward his adversary and inclines his horns sideward toward him. When the two opponents do this at the same time, they are usually standing in head-to-tail position, and then they start slowly circling around each other. When neither of them gives in, they eventually turn from the reverse-parallel position into a frontal position, threaten each other by medial horn presentations, and clash together in a fight.

Even rarer than the sideward angling of horns is the sideward turn of the head in the tommy, but in Grant's bucks it is at least as frequent as the sideward angling of horns and often merges into it, or vice versa. The buck stands broadside to his rival, erects his neck, and turns his head more or less away from his opponent, in the most pronounced but also the rarest of cases at an angle of 90°. In this head-turned-away display, the performer can still watch his opponent from the corner of his eye.

In addition, the head-turning just described can be used as the swing-out movement for a truly magnificent dominance display of Grant's gazelle, head-flagging, to which there is no equivalent behavior in the tommy. Standing most commonly in reverse-parallel position to his opponent and in erect posture, the Grant's buck first turns his head and nose more or less away from but then, by a vehement movement, toward the opponent. In a pronounced performance, he bends his muscular neck somewhat backward so that his white throat flashes up in the sunlight. This dominance display of Grant's gazelle is astonishingly reminiscent of the behavior of a man who sharply jerks his chin toward his shoulder to arrogantly convey to another that he had better leave. And the meaning is precisely the same. The head-flagging of Grant's gazelle simply means: "I am the boss here—move away, you ugly dwarf!"

Reciprocal head-flagging in an encounter of Grant's bucks.

When the receiver is clearly weaker than the sender, he "obeys the order" without any counterdisplay and walks away—according to the usual reverse-parallel position of the two in the direction from which the dominant male came or at least opposite to which he was oriented during the encounter. When the subordinate does not immediately react by withdrawing and/or by assuming a submissive head-low posture, the dominant buck repeats his head-flagging until the weaker one gets the message. In an encounter between at least approximately equally strong bucks, both may repeatedly turn their heads toward each other in the described manner. Sometimes one of them will give in after a while and move off. When this does not happen, both of them eventually change their head-flagging into head-turned-away displays; they will begin circling each other in reverse-parallel position, slowly turn into head-on position, and fight.

By head-flagging and threat-circling, a territorial Grant's buck dislodges trespassing "bachelors" from his territory, and he may also challenge a neighbor in a boundary encounter in this way. Relatively seldom will he use head-flagging in intense herding of females. However, these two expressive displays are frequently seen in all-male groups when the nonterritorial bucks change from one activity to another—from lying to standing, from grazing to moving, etc. The only presupposition for the occurrence of head-flagging is that the other male can at least approximately pass for an opponent to the displaying buck. When the other is hopelessly inferior to him from the beginning—as is, for instance, an adolescent male in an encounter with an adult territorial buck—the youngster is merely being chased away without any display on the part of the strongly superior male. Grant's females also show head-flagging and the other threat and dominance displays, as well as most of the described fighting techniques, although much less frequently than males.

I first studied the territorial behavior of Grant's gazelle in the Togoro plains. On this large mbuga, the following picture emerged: A Grant's buck occupies his territory during the small rains in November/December, and he stays there well into the dry period following the long rains, up to June/July, sometimes even August. During this time, females arrive in small groups, usually only two to five head. The buck herds them, and in most cases he succeeds in keeping them in his territory. In time his permanent harem will grow to about twenty members in this way. If an adolescent male arrives with the females, the territorial buck will usually tolerate him for a couple of days, but sooner or later he will chase him away. Now and then—very seldom as compared to a territorial tommy buck—he will have a boundary encounter with a territorial neighbor, which usually leads to a fight. Trespassing bachelors are expelled by threat and/or dominance displays, as described above. In short, a territorial Grant's buck will not tolerate any conspecific male above the age of a half-grown fawn on his ground on a mbuga in the woodland.

This situation was clear and simple and, on the whole, not very different from that well known to me from Thomson's gazelle. Thus it was rather bewildering for me to watch the Grant's gazelle in the completely open plains at Ngare nanyuki. Here, too, there were all-male groups and groups of females with or without young. However, mixed herds occurred considerably more often than in Togoro, ranging up to a size of four hundred to five hundred head, whereas the biggest mixed herds I ever observed at any mbuga contained about forty members. There were always numerous adult bucks in these mixed herds. However, territorial males and harem groups—even "pseudo-harems" (a territorial male only temporarily visited by a group of females), as occur in the tommy—did not appear to exist in the Ngare nanyuki area.

I observed several particularly strong adult bucks over a longer period of time. Some of them proved to be very dominant over the other males in the herds. Also, they promptly intervened as soon as another buck tried to follow a female, whereas they themselves courted and serviced females. So I came to believe that there was a social hierarchy among the males in a mixed herd and that such a particularly strong buck was the alpha animal* in this herd.

All of this would have been well and good had I not come to know some of these strong and dominant bucks individually. One morning a buck whom I had previously considered to be a highly dominant alpha animal was dominated by an approximately equivalent but by no means essentially stronger buck in the herd, and he gave in without any resistance. The day before, he had still been the most dominant male of the herd. He hardly could have lost his high rank overnight. Although I found it difficult to imagine, I could only surmise that I had made a mistake, that either this buck was not the alpha animal of this herd, or—even worse—I had mixed up the males observed. I did not take my eyes off this buck after that, so that I could not possibly mistake him, and my confusion increased when I saw him behaving again as the undoubtedly most dominant male of the herd later that same day.

A few days later, observation of the buck "Wide horn" completely wiped me out. He was the heaviest buck with the most magnificent horns of all the males around. It was impossible to mistake him, and there never had been the slightest doubt about his alpha position in the herd. One day, however, after a march of the mixed herd over two miles or so, he had a dominance encounter with another male also quite well known to me. I had named this buck "Round horn." I was a bit surprised because I could not remember having seen "Wide horn" and "Round horn" together in the same herd before. I was even more surprised when I saw how "Wide horn," although after some counterdisplays, was expelled from the herd by "Round horn." My confusion increased when, a few days later and in another place, I observed "Wide horn" and "Round horn" again together in a big mixed herd—more or less the same as a few days previously. (The large gazelle herds split and amalga-

*The individual with the highest rank position in a group is called the alpha animal in behavioral sciences.

mate. Thus, size and composition are always changing somewhat.) However, this time "Wide horn" was clearly dominant over "Round horn." I tore the scarce hairs of my scalp and exclaimed in desperation: "In heaven, in the navy, and in Grant's gazelle, nothing is impossible!"

This state so unfavorable to my peace of mind lasted for more than two months. Then the plains burned between the Ngare nanyuki creek and the Barafu kopjes and partly far beyond them. With the exception of Grant's gazelle, there had not been many game animals in the—at this time of the year—completely dry plains anyway. Now the Grant's gazelle had gone, too, except "Wide horn," "Round horn," and five more bucks, two of them also known to me individually and being such doubtful cases with respect to their rank positions in the herds. These bucks stood at great distances from each other on the burned ground, and during the following days it became clear to me that they were territorial. Despite the poor grazing conditions in the burned area, they remained in their territories into the next rainy season. With the first rains and the subsequent growth of fresh green grass, the herds returned to the area. However, now I knew that these seven males were territorial, and I knew the location of their territories, all of which, by the way, were separated by relatively broad belts of "no-man's-land." Now I could see what was going on there, and I began to understand what had earlier been a sealed book to me.

As on a (natural) clearance in the woodland, there are also some territorial Grant's bucks in the completely open plains, and as to size, structure, and other conditions, their territories are, at least in principle, not strikingly different from those on a mbuga. However, a Grant's buck in the open plains cannot keep a permanent harem in his territory, and he cannot prevent other (nonterritorial) males from entering it. Occasionally he may try to do both, but it does not work because the mixed herds are too big here. For instance, the owner of a territory in the open plains may try to stop females from leaving his territory after they have been with him for a few hours, but soon they will cross the boundary despite his efforts and move on, simply because there are many more of them than he can handle. When a mixed herd is inside a territory, it sometimes happens that the territorial male in a truly volcanic outburst of aggression will chase all the nonterritorial bucks out of his territory, thereby separating them from the females. However, they are also too numerous. From all sides, they return to the herd inside the territory, and the owner obviously exhausts himself in constant running and chasing. Therefore, a territorial Grant's buck in the open plains usually tolerates the other bucks in his territory. He repeatedly walks along the file—like a general inspecting his troops—and when each buck whom he passes immediately lowers his head to the ground as in grazing, or actually grazes, or when, upon the slightest indication of a dominance display on the part of the owner of the territory, the other male immediately gives way and withdraws, the territorial buck is content and has no objections to the presence of the bachelors. When one of them follows a female, the territorial buck simply steps between them and takes her away from him. One glance at the impudent rascal is sufficient to make him turn away and move off.

When the mixed herd moves ahead and eventually leaves his territory, there are several possibilities for a territorial Grant's buck. Quite frequently, he simply remains behind, alone in his territory. However, he can also move together with the herd, and once outside his territory, he behaves like any other adult and strong but nonterritorial male in this herd. After a while, he may stop, let the others go, and return alone to his territory. He may, however, also stay with the herd and enter the territory of the next territorial male together with them. In this case, he either is dominated by the owner of this territory like any other male in the mixed herd, or he is challenged by him—be it for not behaving quite as submissively as the other bucks, or be it—more likely, in my opinion—that the territorial neighbors in a given area know each other after a while. There follows a usually long-lasting dominance encounter between the two which may lead to a fight. With or without a fight, eventually the "territorial male without a territory" is repelled by the "right" territorial buck and walks back alone to his territory. In the most complicated case, however, territorial buck A will leave his territory together with a mixed herd, enter the territory of another male, B, together with the others, and be dominated and tolerated there by B as are all the other bucks. When later this herd moves on and leaves B's territory, buck A of course moves with them. But buck B may also leave his territory together with the others, losing his territorial dominance in this way. Now bucks A and B are together in the same herd, both outside their territories, and both behaving like all the other (nonterritorial) adult males. When later, on the same or the next day, in the course of its "circuit," the herd comes back to the territory of buck A, he is on his own ground again, and there he again dominates all the other males, including buck B.

Before I went to sleep in my car, I liked to play the harmonica for fifteen minutes or so. In the evening of that day on which all these things became clear to me for the first time, "Finlandia," my favorite piece of music, sounded particularly nice.

Face to Face with Lions

Quite often people have a rather firm idea about a matter but it has little to do with reality. This may be due to lack of good information and thus be forgiven. Strangely enough, however, some people continue to cling to their imagined opinions even when contradictory evidence is right before their eyes. A good example is the case of the lion, which generally is considered to be "the king of beasts," of yellow color, high-minded, brave, and extremely dangerous. Most of the tourists come to the African national parks with this picture in mind, and ask, "Where are the lions?"

In an area such as Serengeti, lions are by no means rare, and sooner or later a halfway experienced driver will be able to bring his tourists close to them. Usually they will be lying about, looking around, or asleep. This certainly is a peaceful, if not boring, view, but the tourists find it exciting. In spite of the dullness, they remain with the lions and wait for something to happen. In an estimated 95 out of 100 cases, nothing ever does. Sometimes, however, the great event may occur that a lion gets up. Then all the tourists in unison say "Aaaah!" and the cameras start clicking. This is a reasonable thing to do, and I would not advise anybody to hesitate a second, since usually the simba (lion) will go right back to sleep. Only rarely will he walk a few steps before lying down again. When this happens, the busload gets frantic with joy.

The sensible conclusion for the tourists would be that lions—unless they are hunting for prey, which naturally takes up only a fraction of their daily activities, and therefore is seldom watched by short-term visitors to a national park—are bone-idle creatures, and that they—the tourists—have seen nothing particularly royal, wild, or passionate, and usually not much more, on the whole, than they could have seen in domestic cats. Nevertheless, they claim to have witnessed a great thing, seeing these dangerous animals with all their ferocity and rapaciousness.

Even copulation is a rather dull business in lions. In the most interesting but also by far the rarest cases, the lioness may slap at the male's face after the procedure. In addition, with very few exceptions, the scuffles among lions do not reach the verve of a fight among domestic cats. I also could not detect much noble-mindedness when a lion was killing a guinea fowl, a hare, or a gazelle fawn. Even when lions seize larger prey such as adult zebras or wildebeest, they are so much at an advantage with their teeth and claws that their risk is zero. This certainly does not require any special bravery, and I would not be particularly impressed to be told that a prospective adversary was going to fight me with a lion's courage. I would be much more concerned if I were told that he had the dash of a territorial tommy buck. . . .

I can unreservedly agree with only one of the laudatory attributes: the lion is yellow—more or less. And to a certain extent I agree about the danger. A lion is not usually dangerous to humans, but under certain circumstances it may happen. After all, these are large, heavy animals with teeth and claws. Thus, it is not a pleasant experience to find oneself alone, on foot, and unarmed in the wilderness facing a lion at close range. This happened twice to me.

The first encounter occurred at the beginning of my stay in Serengeti. At this time, I regarded the national park as more or less a large zoological garden, and thus I moved about rather easily. While studying the territorial behavior of Thomson's gazelle in the Togoro plains, I soon realized that in order to proceed, I needed a map of the area containing the territories with a rather small representative fraction. Since small landmarks such as termite mounds, single trees, etc. can be of importance to the position of centers and boundaries of territories, they had to be included. A map showing such small details did not exist, however, so I had to make one myself. I had no special instruments for the survey, but I had a compass and a tripod for the movie camera with a spirit level. That was good enough for me; as a soldier with the artillery, I had had thorough training in improvised surveying and cartographical mapping. This survey kept me busy for two weeks, several hours per day. Occasionally I also had to practice the most old-fashioned and primitive survey method, in that I measured a distance by my steps.

The Togoro plains are surrounded by woodland, and in one place the forest jutted out in a wedge shape into the grassland. At first I neglected this section because there were no tommy territories on it. However, upon finishing the survey of the open plains, I decided to include at least part of this not too densely wooded area in the map—if only for completeness' sake. Most of the trees were acacias only ten to fifteen feet high. Approximately at the distance set for my survey, there was a huge tree towering above all the others. I parked my car in the open plains next to a termite mound which I had very exactly surveyed, and I started pacing off the distance to the huge tree, trying to keep a straight line between it and the Land Rover. In walking, it struck me that these acacia trees seemed to be quite suitable for leopards, which were not often but occasionally seen in Togoro. So I decided that in addition to my car and the huge tree, I also should keep an eye on the branches above me. My attention thus was focused forward, backward, and upward.

Having finished step number 220, I saw from the corner of my eye a large, light-brown animal running away from me. It was a lion, and looking around I noticed eight more lions, lying around me in a semicircle about sixty to eighty yards wide, singly or in couples, some of them diagonally behind me. I had walked right into the middle of a resting pride. No sooner had I mumbled a word not quite suitable for print than a lioness resting in front of me got up and approached me in a slightly crouched posture. I do not think that she had immediate hunting intentions. She probably was merely curious and wanted to take a closer look at the strange, two-legged creature that had so suddenly shown up there. On the other hand, her attitude toward me might change for the worse during her close approach. The most stupid thing to do would have been to run away, because this releases the hunting reaction in most predators, and over a distance of more than two hundred yards, it was almost certain that she would catch up with me. So at first I stood motionless. However, when the lioness was within eight yards of me and still coming slowly but continuously closer, I felt that the situation called for some action on my part. I had once seen a lion startled when a bushbuck in the vicinity suddenly uttered its alarm call. This is a short sound similar to the bark of a big dog.

Of all things, now this bushbuck alarm call came to my mind. So I barked at the lioness with a loud "bow"—and she sat down. Constantly facing her, I slowly stepped backward in the direction of my car. When I had gone a few steps, the lioness got up and began to follow slowly. Since I was walking backward, she was still somewhat faster than I, and came closer. Again I yelled "bow!" She sat down, then got up again, followed me, and came closer. After the fourth or fifth "bow," I noticed with relief that I had made it past those lions who in the beginning had been lying behind me, and that all of them were behaving well by remaining in place. It was only this one lioness who would not give up. In this fashion we covered the whole distance to my car, with me "bowing" and her sitting down. When I uttered the last "bow"—because the lady had once more come so very close to me—the door of the car was within my reach. As I got into the vehicle, the lioness remained sitting. Inside the car I had my camera with the telephoto lens, ready for use. I took off the telephoto, put on the standard lens, and took two photos of "my" lioness. Later I found that only the second exposure was well-focused and sharp; the first one, however, I had spoiled by moving.

My second adventure with a lion took place much later, near Banagi. Not far from my bungalow was a flowing stream—that is, it was a "stream" during rainy season. In dry season it was a shallow creek about four to five yards wide. During rainy season we collected the rainwater in large barrels. Used with care, this water supply lasted far into the dry season, but we used it only for drinking and cooking purposes. For everything else water was pumped from said creek and filtered. The only fault in this system was that the pump sometimes quit working, and then it always took weeks to get it fixed. When I returned from my field trips covered with dust and dirt during such a "dry spell," Nyafi, my cook and "jack-of-all-trades," carried water by the bucket up the steep slope from the creek and filled the bathtub with it. Because it was so much trouble for him and the water in the tub was as brown and opaque as in the creek, I told him to spare himself the trouble; I could just as well bathe in the creek.

One day, I again had returned from a field trip of several days and descended to the creek. It was during the dry season, and even in the middle of the creek the water reached only halfway up to my knees. I soaped up well and then lay down flat in the water to rinse off the soap and dirt. At that moment, a lion stepped out of the bushes at the oppo-

Drinking lion.

site bank, descended to the creek, and started drinking—about four yards away from the spot where my head was projecting out of the water. Below the surface my legs were even closer to him. Obviously he had not noticed me, or at least he had not recognized my head as part of a living being. He was a magnificent, maned lion. I was so close that I could count the whiskers around his mouth. I knew I must not stir—and this guy kept drinking and drinking. I cannot tell precisely how long he drank, but it seemed a very long time. Finally he was through and stood there "majestically," looking around, and also at the water and at my head. That was the most unpleasant moment. Apart from the possibility of being detected by him eventually, I was afraid he might lie down right then and there and take a nap. Although it is not a general habit among lions to rest next to the water after drinking, every once in a while one of them may do it. But this lion did the sensible thing. He turned around and disappeared in the direction from which he had come. After that, I pulled myself together and retreated hastily to the house.

As mentioned before, I do not have a great opinion of lions, and I always felt quite happy with my gazelles, who were so much more active and thus more interesting to watch. In both of the depicted cases, however, the lions had my undivided attention.

But perhaps I should not speak so ill of lions. Their roar is impressive, and it belongs to the night in the African plains. I fondly remember the nights when I fell asleep with the lions' roaring in my ears, and the mornings when I awakened to it. Sometimes they were so close to my car that its windows vibrated with their thundering roar. That was truly a magnificent experience.

4 With the Migratory Herds

Most of the game species in the Serengeti National Park shift localities in the course of the year according to the season. Some of them perform regular migrations. Particularly famous in this regard are the (white-bearded) wildebeest, the (Boehm's) zebra, and the Thomson's gazelle. Most spectacular are the movements of the wildebeest, due to their huge numbers. Since the first accurate game count in the Serengeti area, conducted by Bernhard and Michael Grzimek, the wildebeest population obviously has exploded there. In 1958, the Grzimeks counted about 100,000 wildebeest. By 1965, Murray Watson came up with a figure near 300,000, and in 1975 there were more than a million wildebeest in Serengeti. Since then, it appears that they have approximately maintained their numbers. In contrast to the wildebeest in the Ngorongoro Crater, and with the exception of a relatively small population in the far northwestern area of the national park, it can be said with little exaggeration that the wildebeest are constantly on the move in Serengeti. At least twice a year they cross the entire park, moving considerably beyond its boundaries in the southeast during the rainy season and in the north during the dry season.

Alan and Joan Root produced an excellent motion picture on the migration of wildebeest which has been shown on American and European television several times. Although this is a wonderful movie, I have to admit that I watched it with somewhat mixed feelings. In a labor of probably at least two years, the Roots, with an admirable amount of knowledge and efficiency, filmed a multitude of highly interesting scenes and events from the life of wildebeest—which are now compressed into a movie of about one hour. On this basis, to many spectators in America and Europe who are not familiar with Serengeti and the wildebeest, this film easily—too easily—gives the impression that the wildebeest were moving at a wild gallop most of the time, running from the jaws of a lion into those of a crocodile and, after crossing a river in which many of them drowned, directly into the flames of the next bush fire. Of course, all these and many more such dramatic events do happen, but they are distributed over days, weeks, and months. In the long intervals between them, and this means by far most of the time, the wildebeest move along peacefully at a moderate pace, and on the whole their migration is a fairly quiet, although not silent, business. They continually utter their groaning-croaking calls so that the moving herds are, so to speak, enveloped in a cloud of familiar noise.

It is a breathtaking sight when the migratory Serengeti wildebeest approach in almost endless files, or when they disperse in grazing from horizon to horizon. When accompanying the moving herds in my Land Rover in the beginning of the rainy season when the air was prickling like champagne, I always felt a true flush of wanderlust, of unlimited roaming across endless plains.

As mentioned before, the zebra in Serengeti are also migrants. For a long time, we had a rather incorrect view of their social organization. Plains zebra were known to occur in relatively small as well as in very large herds. The big herds were considered particularly typical, and it was presumed that each of them was led by a strong stallion, "the master of the herd." So it came as a surprise when my colleague Hans Klingel found out that these big herds are rather loose associations, and that nothing like a "leader stallion" exists in them. The basic social units are the small groups into which the big herds split up after some time. In part, these small groups are "harem" groups consisting of one adult stallion and two to six mares with their offspring. These harems remain very stable over the years in plains zebra; they roam the country or may stay in a given area for a while, however, without any indication of territorial behavior. Young stallions leave the harem groups in which they were born at between one to four years of age, usually without being chased away by the fathers. They form all-male groups which are as small as, often even somewhat smaller than, the harems. Later, such a "bachelor" may try to separate young mares from harem groups—one may safely say that he tries to steal them. Usually, one bachelor alone will not be very successful, owing to the vigilance of the "boss" of a harem. However, when several bachelors harass the same harem group simultaneously, at least one of them may eventually prove successful, and he may start forming his own harem in this way. Thus, the two basic units in plains zebra are the very stable harem groups and the all-male groups, the members of which may also remain together for several years. Female groups, so characteristic of many gregarious antelope species, are not found in plains zebra. When zebra form herds of hundreds, thousands, and tens of thousands during migratory periods, these are composed of many harem and all-male groups which often are still recognizable as subunits within the large herds. So much for the results of Hans Klingel's study. More I will not say here about wildebeest and zebra, since both tommy bucks and scientists are well advised to respect "territorial rights" and not venture too far into the territories of their neighbors or colleagues, respectively.

I am back on my own territory when talking about Thomson's gazelle. The British ecologist Brooks reported that at the beginning of a dry season, the immigration of Thomson's gazelle into the woodland follows that of zebra and wildebeest, whereas at the beginning of a rainy season, their emigration toward the open plains precedes that of wildebeest and zebra. Based on these findings, the scientists and the game wardens in the national park were of the opinion that of all the game species there, Thomson's gazelle stayed the longest in the open country (a point which appeared to be important with respect to the availability of prey for some nonmigratory predators). As I soon found out, however, this view was not completely correct. Some of the Grant's gazelle who are in the open plains during rainy season migrate and leave these areas more or less together with the tommies. At least as many Grant's gazelle, however, do not participate in this migration but stay in the plains during dry season. Thus, not Thomson's but Grant's gazelle are the last ones in the open country. Only when the drought peaks in August/September do they, too, start moving toward the woodland—so slowly, however, that they often arrive at the edge of the forest only a short time before the beginning of the next rains—and with the onset of the rains, they turn around and emigrate back to the open plains.

Roughly speaking, the Thomson's gazelle leave the woodland in the north and northwest of the Serengeti National Park with the onset of the small rains in October/November, and they move far out to the southeast. Provided that the rainy and dry seasons occur at least approximately according to the calendar (which does not always hold true in East Africa), the tommies remain in the open plains and move about in the short-grass areas between Lake Ndutu (formerly Lake Lagarja), in the far southeastern "corner" of the park, and the Lemuta hill, about forty miles, as the crow flies, northeast of this lake, outside the park's boundary, where they stay until the long rains end in May. If the small rains have been meager so that the open plains are bone-dry again by December or January, and/or when the long rains do not begin in March, as they should according to the calendar, the tommies may prematurely return to the edge of the woodland. As soon as the delayed rainfalls take place, they move out into the open plains once more.

After the end of the long rains, the tommies definitely start their migration toward the woodland—except for a very few individuals, usually old males, who do not take part. The herds cross the central high-grass areas of Serengeti in a northwesterly direction, and—again provided that the seasons change according to the calendar—they arrive at the edge of the woodland at the end of May or the beginning of June. There they gather along the rivers—which at this time are already partly dried up—particularly the Seronera and the Mbalangeti, and, following the river beds for some distance, they move deeper into the woodland.

With this, their predictable migration is at its end. What happens later on depends on the extent of the drought, on the distribution of short grass and high grass on the mbuga in the forest area, on the extent of burning, and on the rare, locally very limited rain showers which may occur during dry season. These conditions vary from one year to the other, and correspondingly, the movements of the gazelles within the woodland are somewhat different each year. Extensive fires in the forest may cause a food shortage for the animals, especially in the last third of the dry season.

Except for some herds which remain on a large mbuga (such as the Togoro plains) in the woodland, the Thomson's gazelle gather at the edge of the forest area again during the last days or weeks of the dry season, waiting there for the first rains to go down in the plains. As soon as this happens, the emigration commences as if a starting shot had been fired. Particularly in the beginning of a rainy season, the first showers are often strictly local, sometimes covering not much more than one square mile. After that, you can almost see the fresh green grass growing there. Of course, such a place may be miles away from the area where the gazelles are, yet they usually arrive there after a very short time. How do they do it? As far as I know, there is no exact proof for it, but I can venture a quite fair guess. When driving through seemingly still dry country at the very beginning of a rainy season, it happened several times that I suddenly felt a taste of humidity on my lips. In all the cases in which I could check the matter, it turned out that there had been a rain shower in this area a short while before. As is well known, the human ability to perceive odors and tastes is very limited as compared to that of animals such as ungulates. When a man can perceive signs of humidity in otherwise dry surroundings, the assumption does not appear too far-fetched that animals, with their keener senses, can do it even better and over considerably longer distances.

If a tommy could tell us what "heaven on earth" would look like for him, my guess is that he would wish for—besides the presence of conspecifics and the absence of predators, parasites, and diseases—a wide, open land with warm days and cooler nights, a few shade trees here and there, solid, dry ground covered with green short grass, and a constant wind neither too strong nor too soft. Let us take a closer look at these "heavenly" conditions, since most of them are not without relevance to migratory behavior.

When exposed to heavy rain, Thomson's gazelle turn their hindquarters toward the wind and rain, slightly arch their backs, and somewhat pull in their necks. In this "attitude of discomfort" they stand in the downpour "waiting for better times." All of this they have in common with quite a number of other species of open plains animals. Somewhat more uncommon is the habit of Thomson's gazelle to lie down at the beginning of a rain. In principle, this may happen in some other species as well, for instance, in Grant's gazelle, but in none of them is it as obligatory as in the tommy. Also, in the other species, there are usually only some individuals in a herd who behave this way, while others remain standing. By contrast, entire herds bed down in Thomson's gazelle. A further response which is exclusive to tommies is fleeing from a sudden downpour. It does not occur as often as lying down but is still frequent enough. Thus, with the onset of a rain, a tommy herd will first flee with the wind from behind, running fifty to one hundred yards. Then the animals will stop and lie down. When the rain continues for a longer while, they will rise one after the other and wait standing for it to end. Usually, such a rainfall does not last much longer than two hours in East Africa, often less than one hour. During this relatively short time, however, it rains cats and dogs. As mentioned before, the plains rise and fall in soft "waves." During the rainy season, the grass grows first in the depressions, and in the beginning of the dry season, it remains green there longer than on the elevations. Therefore, the tommies—as well as some other game animals—often graze in these low spots. Consequently, their relatively short flight at the beginning of a heavy rain brings them out of the depression and up to the elevation, and this is probably the "purpose" behind this behavior. Given the hard African laterite soil, the water may soon stand one to two inches high in a depression during such a rain. Also, when it later soaks into the soil, the tommies avoid—here as otherwise—the moist, softened ground. A tommy does not like to get its feet wet!

When studying Thomson's gazelle, it is particularly pleasant for the observer that he is relatively little bothered by insects. (Some molestation by insects can hardly be avoided in any kind of field work in Africa.) In any case, it is nothing compared to what one may experience when watching certain other game species. For example, on the way from Banagi to the Togoro plains, there were a few rocks near the road in the woodland. When passing along, I almost regularly saw klipspringer there. One day I wished to take a closer look at them. While driving through the woodland, I had always been bitten by tsetse flies. I was used to it and did not take it "tragically," not least because I knew that tsetse flies only very exceptionally transfer sleeping sickness to humans in the Serengeti area. However, my experience with them while parked in my car observing the klipspringer was so horrible that I stopped the observation after three hours with wild cursing.

There are no tsetse flies in the open plains. However, the herds of wildebeest, as well as the cattle of the Masai and the Masai themselves, are constantly surrounded by small flies which do not bite but are very annoying due to their incredible numbers. The wildebeest, like many other ungulates, swish their tails, shake their heads, etc. in an attempt to defend themselves against flies. On the whole, however, they accept this pest rather stoically. The situation is different in Thomson's gazelle. When, during the migration in the woodland, a tommy is heavily bothered by insects, he stamps with fore- and hindlegs as if the ground had become hot under his feet, grooms and scratches frantically, and, after quivering of his flank (the same sequence of behavior as in a conflict situation), he takes off at a gallop. Occasionally a whole herd may do this. This sensitivity toward insects is probably one of the reasons why tommies always leave the woodland as soon as possible and readily return to the open plains, where there are not only fewer insects but also a pretty steady wind to blow them away. In the densely crowded wildebeest herds, of course, the flies can stay despite the wind. When these herds arrive in an area where gazelles are present, the tommies will promptly move away. Perhaps a few tommy bucks may at first try to keep their territories; however, when the wildebeest remain in the area for several hours, they too give up and disappear. I am not sure whether this is primarily a reaction to the masses of the wildebeest per se or to the masses of flies they bring with them. (I have the impression that the number of these Masai flies has increased with the increase in the wildebeest population in Serengeti.)

Although I never conducted a special study on the food plants eaten by gazelles, it struck me how strongly they prefer green grasses, green herbs, and green leaves, although they of course will eat dried-up yellow or grayish plants when there is nothing else left during dry season. The availability of a bunch of green grass in otherwise dry surroundings presents one of the extremely rare situations in which gazelles may fight over food in the wild. It may even cause an interspecific interaction; for instance, an adult tommy buck may fight a female or a half-grown Grant's gazelle over such a morsel. In addition to these aggressive interactions, the gazelles' preference for green plants becomes particularly striking in two situations.

When the drought approaches its culmination point, the vegetation in the woodland is eventually affected as well. As long as there is still some water in a river bed here and there, the grasses and bushes along its fringes remain green somewhat longer than in other places. Then the tommies come to the river as a daily routine not only to drink but also to eat extensively from the green plants there. Sometimes—in Grant's gazelle, usually—they do not drink at all but exclusively graze the green grass at the river bank.

Secondly, when in the beginning of a dry season the grass becomes yellow and grayish in the open plains, it often remains green for a while along the edges of roads and car tracks and, above all, in the immediate vicinity of termite mounds. Thomson's gazelle gather at such locations and graze intensely. To a human observer, the gathering along the road and car tracks is often quite welcome in that he does not need to drive so much across the country in order to meet the animals. The grazing around termite mounds, however, I often wished to hell. Frequently, almost always in Serengeti, aardvarks establish their subterranean dens in the vicinity of termite mounds. Later, when the original "constructors" have left, some of these dens are occupied by hyenas who enlarge the entrances. Consequently, there frequently are holes in the ground near a termite mound in which a car can get easily and badly stuck.

When, in the evening, banded mongooses shoot up from the airshafts to the top of a termite mound, the tommies grazing around it almost regularly take to flight.

Grazing near a termite mound can sometimes lead to a quite amusing but also very instructive scene. Each termite mound has several holes about the size of a child's head. These are the openings of airshafts often running more or less vertically down to its depth. Banded mongooses inhabit these shafts. They spend the hottest hours of the day inside but come to the surface later. They pop up abruptly from the openings of the airshafts, sit on top of the termite mound, and look around before they go in search of food. When there are Thomson's gazelle grazing nearby at this moment, they flee as if a bomb had exploded among them. After running for a hundred yards or so, they stop and look back. When they see that there are only the little mongooses, they calm down and start grazing and approaching the termite mound again.

Likewise, I found in some other situations that a tommy will immediately run as soon as something occurs suddenly close by, regardless of what it may be. Certainly, this is a very useful reaction. Predators of gazelles are numerous in the East African plains. Thus, anything that suddenly and surprisingly shows up near a gazelle could be a predator, and even a moment's hesitation can prove disastrous to the gazelle. Therefore, it pays for a tommy to run immediately in such a situation and not lose any time by investigating the object which has shown up there so suddenly. If the flight turns out to be for nothing, as in the case of the banded mongooses, who do no harm to a gazelle, well, this is a disadvantage which has to be put up with.

In a tommy's "heaven on earth," however, the grass not only should be green, it also should be short. Like apparently quite a number of mammals living in open country, Thomson's gazelle rely strongly on visual orientation. In the woodland and in high grass, this visual orientation may easily be lost. In my opinion, this is the most important reason why Thomson's gazelle so strikingly prefer the short-grass areas and enter dense forests and high-grass areas only when they cannot avoid it, which means predominantly when they have to pass through in the course of their migrations. The wide view in the short-grass plains may also offer some advantage with respect to the avoidance of predators. Above all, however, it is very important for visual contact with conspecifics and for orientation toward landmarks.

In the Serengeti area, such landmarks include kopjes, single trees, single and relatively high hills, and chains of hills. The latter may sometimes be pretty far out at the horizon. When I accompanied the migratory tommy herds, I soon realized that in most cases they had such a landmark in sight and were heading toward it. Frequently, they do not walk directly up to the object. For instance, at first a kopje may be the only landmark visible to them, and they move toward it. After advancing in this direction for a while, if they spot a second kopje (or any other prominent object) situated ahead in the general direction of their move but somewhat sideways from the first kopje, they may change their course in the direction of the new object. So they migrate from one landmark to the next. Also, the river beds, with or without water, play a great role as "guiding lines" which the gazelles may follow for miles. Finally, a conspecific may become a "landmark." At least as long as it is on its feet, any animal the size of a gazelle (or taller) affects the vision as a "towering" mark in the short-grass plains. Obviously, tommies can distinguish at a distance of at least one mile whether there are other animals or some of their own "people"—possibly, or even probably, the black stripe at the flank and/or the continuously wagging black tail of Thomson's gazelle may be helpful in this species identification. Grazing or standing conspecifics are more attractive than resting ones, moving groups are more attractive than those that are stationary, and large herds are more attractive than small ones. So a relatively small group may deviate from its course at a right angle or even turn and move back in the direction from which it had come when a large herd shows up there. In this way, the herds flow together in times of migration and may occasionally form concentrations of thousands of animals.

The importance of visual orientation to Thomson's gazelle becomes particularly clear when they have to cross a high-grass area on their migration from the short-grass plains toward the woodland. When the grass is higher than the tommies—which does not take too much, since a tommy has a shoulder height of only a little more than two feet, and grass as high as wheat is enough to make one disappear—their visual orientation gets lost. Thus they often hesitate before entering a high-grass area. When the first animals, usually females, of a mixed migratory herd arrive at the edge of a high-grass region, they almost always stop. Since more animals continue to arrive from behind, they soon gather into a tight bunch, in which adult males are also increasingly involved. These try to push the others ahead with all their might—one of the very few situations in which a tommy buck may threaten a female as if she were another male, by presenting his horns toward her. The result is at first a great confusion, with the animals rushing to and fro. Sooner or later, however, the first animals, usually females again, "dive" into the high grass at a gallop. The others follow, and in persistent gallop, they try to cross the high-grass area in the adopted direction. If the high grass covers a large area, the herd will split up into small groups running in every direction, with most of them finally returning to the place from which they started. There they stand, craning their necks to see over the high-grass "ocean" toward a far-out landmark, for instance, a high hill at the horizon. Again the adult bucks heavily push the others, and again these "dive" into the high grass at a gallop. Eventually they always make it, but it may take them many trials.

Whether the availability of drinking water is the most important factor for the migration of Thomson's gazelle from the plains into the woodland, or whether their movements primarily follow certain food plants, is a question I left with pleasure to the ecologists more competent to solve such problems. I was more interested in two other questions; however, I have to admit that I did not get far with them.

What causes the migration of the gazelles from the woodland into the open plains is relatively clear: the first rains in the plains. The opposite move, from the plains to the woodland, is linked, of course, to the increasingly dry conditions which, in the beginning of a dry season, take effect in the open country earlier than in the forest. Nevertheless, the question remains as to what factors actually and concretely initiate it. Several times at the beginning of a dry season, I was in an area optimally populated by tommies. The short grass was not marvelously green anymore but it still had a greenish shine, and the animals took it quite readily. The territorial bucks were engaging in boundary encounters, and they chased bachelors and herded females. All-male groups and female groups moved through the area according to their daily "circuits," and so on. In short, everything looked as if the gazelles would stay there for at least a few more days—and with this conviction I went to sleep in the evening. When I woke up the next morning, however, the gazelles around me were in full migration. The grass looked as it had the day before, and when I measured the temperature, the humidity of the air, the velocity and the direction of the wind, etc., I did not find any striking differences as compared to the previous days. However, the great move had started. I do not know what sets it in motion in such a case.

Furthermore, as I described above, the orientation of the migratory herds toward landmarks is often clear and obvious. However, this seems to be a finer orientation. Thus, the question remains: What gives the general direction to the animals? I became aware that standing and, even more so, lying gazelles commonly orient themselves with their backs toward the sun and/or toward the wind. About nine o'clock A.M., the time of the obligatory morning rest, it is quite a strange sight to see hundreds or sometimes even thousands of gazelles bedded down in the open plains, all of them facing in the same direction. Thus, it is imaginable that the sun and/or the wind—which in Serengeti blows rather steadily from the east or southeast, and usually with quite remarkable velocity in the morning hours—may also be used for orientation during migration. Unfortunately, I cannot provide any hard proof for it.

I had better luck with the observation of individual animals during migration. Such observations are very necessary. It is by no means true that the gazelles are continually on the move during a migratory period. During daytime, phases of busy marching alternate with at least equally long pauses during which the animals graze or rest. Before and after dusk, there is usually another significant move, but later during the night, the tommies commonly do not migrate more than one to two miles, which is about as much as a female herd or a bachelor group may move at night in a stationary period. Also, in the dry-season migration, the tommy herds do not approach the woodland in a straight line. They frequently move on a zigzag course, occasionally even walking back in the direction from which they had come, over some distance. Consequently, when I measured on a map the distance, as the crow flies, which a tommy had covered from early morning to late evening on his way toward the woodland, it never turned out to be more than ten miles at maximum, and usually only five miles or even less—although he actually may have walked a much longer distance. Finally, there is much splitting and amalgamating in the herds, as there is among gazelles during more or less stationary periods; during migration, however, it occurs continually. A group will separate and remain grazing at one place while the others move ahead, then a new group will join the herd, and so it goes all the time. It can only be said that the tommies migrate in herds, but it is almost impossible to speak of a definite herd. Taking all these aspects together, it only remains to select individual animals and record their activities when one wishes to learn something about details of behavior during migration.

I had conducted a similar study with territorial tommy bucks and with bachelors in all-male groups because I wanted to quantitatively compare the occurrence of certain behavior patterns—such as threat and courtship displays, fighting, marking, etc.—in the different social classes. For this purpose, I had watched ten territorial males and ten adult bucks in all-male groups, each of them individually, from dawn to dusk, recording what they did and how often they did it. All the territorial bucks were "old friends of mine" who remained neatly in their territories. This investigation was a true pleasure. Likewise the bachelors selected were all known to me from previous observations. However, it was a little more difficult to keep constant track of each of them among the other males and on their daily "circuits" within their home ranges, which, of course, were larger than the territories. In a similar way, I now had to record the behavior of migratory bucks. However, I had not individually known one of them before, they moved over considerable distances, and they often were in a crowd. This was hell.

As part of this study, I also wished to obtain records on the activities of ten adult migratory males from dawn to dusk. In the very first light in the morning, with binoculars at my eyes, I searched through the herds, which at this time were usually still standing and grazing. I was looking for the strongest male with the longest horns I could detect, in the hope that it might be relatively easy to keep track of such a male and to distinguish him from the others during the coming migration. Once I had found a buck who appeared to be suitable, I had to keep my eyes on him for the next twelve and a half hours. Shamefully often this did not work out. Again and again it happened that I lost track of "my" male in the crowd or when he was crossing a high-grass area, etc. Or I broke through into a hyena den, or I ran up against a rock hidden in high grass, since, of course, I could not keep my eyes on the buck and at the same time carefully watch where I was driving. By the time I got the vehicle going again after such an incident, "my" buck had disappeared into the "vast plains of Africa." In all these cases, the work of the whole day was spoiled—at least as far as the records on the behavior of that individual male were concerned. When I add up all the time it took me to get ten complete and irreproachable records on the daily activities of individual migratory tommy bucks, it amounts to one and a half months. But the trouble was not in vain. Beside data on the distances traveled, on the distribution of marching, resting, and grazing periods over the day, etc., these investigations revealed a few behavioral "delicacies."

All adult tommy males, regardless of their social status, use the secretion of their preorbital glands to mark grass stems, small branches, long thorns, and similar objects. However, territorial bucks do it significantly more frequently than bachelors in the (stationary) all-male groups. The migratory bucks fall quantitatively in about the middle between them. It could be presumed that they mark the migration routes—and, in a sense, this is probably not totally inaccurate, but it is by no means as simple and clear as it may theoretically appear to be. The fact is that these bucks mark most frequently upon departing from a place where they have rested or grazed for a while. The busier their march, the rarer become their marking actions. In crossing high-grass areas—where their visual orientation may easily be lost, and therefore, according to human considerations, some additional means, such as olfactory marks, might be quite helpful—they do not mark at all, as far as I could figure.

Under quantitative aspects, I found all behavior patterns linked to social life—such as courtship displays, fighting, chasing, secretion marking, urination/defecation in a sequence, etc.—to be significantly diminished among the bachelors in the (stationary) all-male groups as compared to territorial bucks. Some behaviors, as, for instance, (true) copulation, were not seen even once in them throughout the entire study. However, there is one exception to the rule: threat displays by presenting the horns are significantly more frequent among bachelors than among the owners of territories. In territorial bucks, such threat displays are restricted to boundary encounters between neighbors and to expelling trespassing bachelors—if the latter are not simply chased away by the owner of a territory without any previous threat. Provided a territory is located in an area well populated by Thomson's gazelle, such events will predictably occur in the course of a day, however, usually in relatively small numbers. By contrast, the bachelors in all-male groups first may have "arguments" over the individual distances among them. Furthermore, and above all, they synchronize the group activities by threat displays. When, for instance, an all-male group has rested for a while and some members have gotten back to their feet again while

SYNCHRONIZATION OF GROUP ACTIVITIES BY AGGRESSIVE INTERACTIONS

Above: In an all-male group of Thomson's gazelle, some of the members are already on their feet while others are still lying at the end of a resting period. One of the standing bucks "takes offense" at a resting companion and threatens him by presenting his horns . . . *Below:* . . . until he rises.

others are still lying down, at least some of the standing or moving bucks will start presenting their horns toward resting companions and thereby make them rise. In a somewhat larger herd, a male may "pester" several resting comrades, one after the other, and since gazelles have several resting periods in the course of one day, an individual buck may carry out quite a number of threat acts during such activity changes. Furthermore, the bachelors in all-male groups may threaten each other when "voting" on the marching direction. For instance, when they are departing from a place where they have rested before, it is not unusual for some members of the group to first walk in different, even opposite, directions. Upon meeting each other, they threaten each other until all of them move in the same direction. Finally, when marching in a file, a male will threaten the buck in front of him when the latter slows down or deviates from the course of the rest of the herd. In short, the bachelors in the all-male groups have more motivation for threat displays, and situations in which threat displays may easily show up occur more frequently among them in the course of a day than in territorial bucks. In all these encounters, the older and stronger bucks display much more unconstrainedly, more effectively, and also more often than the younger males, who more frequently receive and obey the threats. In a moving all-male group with bucks of different ages, the result is that the youngsters, threatened and driven by the older males, march at the front of the file.

All of these points are equally true for the tommy bucks in the migratory herds. They, too, synchronize and coordinate the group activities by threatening each other, they speed each other up during a move, etc., so that they present their horns about as frequently as the bachelors in all-male groups. However, there are also many females in the mixed migratory herds. Only in very special and, on the whole, rare situations will a tommy buck threaten a female by presenting his horns. In most situations where he would use this threat toward a male conspecific, he shows gestures or postures toward a female which also occur in courtship and in territorial herding, predominantly the neck-stretch (head and neck stretched forward on body level) and the nose-up (neck erect, nose lifted upward). Mounting may sometimes be seen as well, although much more rarely than the aforementioned "courtship displays," which in frequency come close to the horn threats in migratory males. (Among the stationary bachelors, they are almost nonexistent.) Whether and to what extent these "courtship displays" may still be sexually motivated in a migratory buck (his marching behind a female resembles the male's following a female in the mating ritual, and therefore may possibly lead to some kind of a "transitional action") is a question which cannot be decided by observation—and, unfortunately, we cannot ask a tommy buck what he "means" by it. In any case, the behavioral inventory of a tommy buck toward females is largely different from that toward other males. (In the relatively closely related Grant's gazelle, this difference is not nearly as clear and pronounced as in the tommy.) During migration, a male's "courtship displays," the neck-stretch and the nose-up, serve the same functions and have the same effects upon females as do his threat displays, the presentation of horns, upon other males.

The pushing activities of the adult tommy males automatically bring the females to the head of a moving mixed herd. In a relatively small migratory herd, all the females may walk in front of the males. In a large herd, several females move to the front, usually followed by an intensely driving adult buck. Then come the majority of the females and subadult animals of both sexes interspersed with a few adult males at irregular intervals. Finally, a group of adult bucks marches at the herd's rear. Thus, it cannot generally be said that females "lead" the migratory tommy herds. Most of them simply march at the head because they are the most intensely driven animals in a mixed herd. Sometimes, however, a female with an almost completely white forehead (i.e., apparently an old and presumably experienced individual) is seen walking as the first one at the head of the file. In such a case, it may be justifiable to speak of this female as the herd's leader. In a very few, apparently rather difficult, situations, a male with an almost completely white forehead may also temporarily take the lead of a mixed herd. This was only exceptionally but clearly observed.

With their pushing threat displays toward other males and their pushing "courtship displays" toward females, the adult bucks are the "motors" of migration in Thomson's gazelle. When the herds eventually arrive in an area with favorable conditions, one adult male after the other leaves the herds and becomes territorial. When numerous neighboring territories subsequently form a mosaic field, the owners of these territories chase the other (not yet territorial) males out of the mixed herds when they move through this area. In these ways the herds lose their "motors," and the migration comes more or less to a standstill. Quite impressively, the territorial behavior occurs as the counterpart of the migratory behavior in such cases.

Of course, this is also valid in the opposite situation. Some tommy bucks retain their territories when the other gazelles are already migrating. Then it may happen that the herds move along such a territory or even right through it, and the owner exhausts himself in futile attempts to stop and herd the females inside his territory and to keep the males out of it. Eventually he will give up and lie down with his back to the wanderers. "He cannot look at it any longer." More and more herds pass along, and sooner or later he is "washed away" by them.

Although Thomson's gazelle are much smaller than wildebeest and zebra, they maintain relatively larger individual distances* among them. Except in fighting, in copulation, and in mothers' nursing and cleaning of their young, free-ranging gazelles hardly ever touch each other. At minimum, they keep a distance of a few inches from each other. Moreover, their individual distances vary somewhat with activities. On the average, the distances among gazelles during a rest and during a move are about equally small; however, they are significantly enlarged in grazing. Then, again on the average, the distances between males are larger than those between females. (The latter has a few quite interesting consequences; for example, when two territorial bucks step backward in frontal orientation in a grazing ritual after a clash, they resume the usual grazing distance among males.)

Particularly striking, as compared to wildebeest and zebra, are the relatively larger individual distances maintained by gazelles in concentrations. In Thomson's gazelle, concentrations of thousands and tens of thousands of animals last no longer than one or two days, often not even half a day. Although they occur only during migratory periods, they generally (i.e., also in the other species) are much more prominent during the grazing and resting pauses than in actual migration. While wildebeest and/or zebra may easily form a rather united crowd, gazelles remain relatively scattered, even in a concentration. Seen from a distance, the edge of a tommy concentration also may at first give the impression of a wall. However, when I wanted to drive right into the middle of such a concentration, I regularly had a funny experience. In close approach, I had only single animals or small groups of two to five in front of my car. Thus, I presumed I was not yet in the true concentration, and I drove ahead, from one animal to the next and from one little group to the other, until I was through, eventually, and had the concentration behind me.

---

*Hediger, who was the first scientist to discover the important fact of individual distances in animals and who coined this term, defined it to be the minimum distance to which two conspecifics can approach each other without evoking agonistic reactions. At least for certain purposes, however, I prefer to consider the individual distance as the average distance (with its standard deviation) observable among group members. When the records from many groups are added together, the resulting average may be said to represent the individual distance typical of a species.

Even in a concentration, Thomson's gazelle maintain relatively large individual distances.

In contrast to wildebeest, gazelles migrate silently. Only when I was very close to a large, relatively fast-moving herd could I hear a soft noise as if a gentle wind were blowing over a field of wheat. This was caused by the hundreds or thousands of hooves and legs touching and gliding through the grasses. It sometimes is remarkable how deeply small details can impress us without our becoming aware of it when it happens. At that time, I did not attach any significance to the noise of the gazelles' legs moving through the grass. However, after years of being far away from East Africa, when I think back on the migratory tommy herds in the Serengeti plains, I long to hear this gentle blowing sound once again.

5 Under Southern Stars

When I began my study in the Serengeti National Park in 1965, not much was known about the twenty-four-hour activities of African game animals. When reading the relatively few reports available at that time, I hardly could suppress the ugly suspicion that they sometimes reflected more the observer's schedule than that of the animals. For example, in these reports, the animals' activities usually were strongly reduced between twelve and two o'clock P.M.—just the time when a good citizen takes his lunch and a nap; and at night the animals were at rest—just like a man after a good day's work. Not far from our research station, one of the very few, small herds of roan antelope in the Serengeti National Park used to spend two to three months a year at the foot of the Banagi hill. Using a good spotting scope, I could watch them there from the window of my house, and thus, without conducting a special investigation, I found that they fed in the early morning and in the late afternoon or the early evening, respectively, but rested between these two activity periods—except at noon, when they regularly came to the Orangi River near the research station to drink; and this was by far the longest move they made in the course of their daily routine at that time. Also, I soon learned that the best time to observe the mating rituals of Thomson's gazelle in the Togoro plains was around midday, at which time courtship often was particularly frequent and intense. Only around two P.M., thus more in the early afternoon than at noon, did their activity begin to dwindle quite drastically, especially on hot days.

Having found that the "noon rest" was a somewhat dubious matter at least in certain species, I was no longer inclined to take the "night rest" for granted either, and I decided to check it in "my" animals. This led to quite a number of nighttime observations. Thomson's gazelle bed down half an hour to one hour after nightfall, which means about 7:30 P.M. Between 9 and 10 P.M., they often are on their feet, but not regularly so. When it occurs, this activity period does not last long, and then the tommies rest again. To me, this activity seemed to be more a disturbance in a genuine resting period of these animals, possibly or even probably caused by the behavior of certain predators, such as lions and hyenas, who apparently have an activity bout at that time. Of course, such a "secondary" activity induced by the activities of predators is a normal part of life for a prey species in the wild. After 11:30 P.M. the rest is over, and the gazelles are active for at least one hour on dark nights. On bright moonlit nights, they may be active up to about 4 in the morning. Then they rest until daybreak.

I do not know whether infrared light and starlight binoculars were on the market yet in 1965/66. In any case, I had none. On bright moonlit African nights, my binoculars were sufficient for observing single events. For a more systematic study of nighttime activities, which, of course, also had to include dark nights without moonlight, I used a movable searchlight charged by the battery of my car. When I searched through the surroundings with it, the eyes of the gazelles flashed up with a yellowish-green shine, and I could at least recognize whether they were standing, resting, grazing, or moving.

I soon learned that not all the game species reacted to the searchlight in the same way. For instance, zebra immediately ran when met by the beam. If they had been among the gazelles, the latter ran, too, and my nighttime observation was over. Fortunately, the Thomson's gazelle tolerated the beam much better, but I also had to gain some experience with them before it worked. Completely undisturbed, as I wished and needed it in this study, they continued their activities only when I left the beam on them for a very few seconds—comparable to a flash of lightning, which does not disturb them either. However, when I left the beam on them somewhat longer, they interrupted their activity, looked out toward the source of the light, became restless, and eventually ran toward the light at a gallop. Of course, I did not know this early on, and the first time I tried to use the searchlight, I left the beam on them too long. A whole herd came running at full speed toward my car; many of the gazelles missed it by only inches, and one of them crashed into the standing car. Later on, I switched on the searchlight only every quarter of an hour and moved it around in a circle, leaving the beam on each animal only for the moment which was necessary to recognize its activity. Simultaneously, I spoke into the tape recorder "Standing—resting—grazing—grazing—moving," and so on. Based on these records, I could later figure the percentages for individual activities with respect to the total number of animals met in each "transit."

This worked quite well. Sometimes, however, a peculiarity of "my" gazelles gave me trouble. Particularly after midnight but occasionally somewhat earlier, the large herds tended to split into small groups. (After daybreak they united again.) About eight o'clock in the evening, I might still be surrounded by more than one hundred tommies, but during the course of the night there might gradually be fewer and fewer of them until in the morning, about four o'clock, I was left with only five or six. In the beginning, I drove around in the dark in such a case and tried to find another, larger group. However, this proved inopportune. I only made the animals spooky in this way. Thus, all I could do was stay the whole night at that place where I had been at nightfall. When I was unlucky and the number of gazelles visible around me became too small for a later statistical evaluation, the watch had been in vain, and I had to repeat it. By the way, such a watch was truly "in vain" only with respect to the investigation of the twenty-four-hour periodicity of the tommies. Otherwise, I did not regret a single night which I stayed awake in Serengeti. Even after I had collected a number of records on the activities of gazelles which appeared to be sufficient for my purposes, I continued to conduct nighttime observations from time to time. I learned something about the African night sky, and I enjoyed looking up to the southern cross when it was visible, which is not true all year round in these latitudes. Above all, I was endlessly fascinated by the nocturnal life of the animals.

This began with "my" Thomson's gazelle. Of course, at night they did not show any different behavior than during the daylight hours. However, "It is night; now all the fountains speak louder,"* as the philosopher Nietzsche poetically put it. There were no fountains in Serengeti; however, here, too, sounds and noises were more prominent at night.

*"Nacht ist es: nun reden lauter alle springenden Brunnen . . . " (F. Nietzsche, *Also sprach Zarathustra*)

They were often my first indication that something was going on in the surroundings. For instance, I suddenly heard a clattering noise somewhere in the night, and when I looked with binoculars in its direction, I saw the moonlight shining on the horns of two fighting tommy bucks. Or, I heard a soft bubbling sound nearby—the driving call of a tommy male, which he utters by nose. When I looked toward it, the silhouettes of the "wedding couple" showed up in the dark, and the buck's neck and head were set off against the night sky in his nose-up display.

Furthermore, there were some species, such as lions and hyenas, which could also be seen during the daytime but usually only at rest, whereas they "came to life" at night. The only copulation of kori bustard I ever saw took place at night. Finally, there were animals which never or only rarely showed up during the day but were strongly nocturnal, such as porcupine or several species of genets.

During one of the first nights I spent in the field, the beam of the searchlight met the top of a tree by chance, and there, two eyes flared up orange-red. At that time, I had no idea what animal it might be. Sometime later, I was driving at night along a small path in the woodland. Suddenly, two eyes of the same color as in that tree flared up in front of my car, only this time the animal was on the ground—and then a bush baby (a galago, related to the lemurs; it gets the name "bush baby" from its cry, which sounds something like that of a very bad-tempered baby) jumped high up from the ground into the branches of a small tree. Very probably, the mysterious animal on top of the tree on that former night had also been a bush baby.

Another time, the beam of the searchlight caught several eyes flashing up in a yellowish-green shine like those of gazelles. However, the carriers of these eyes moved in an arc-like fashion and closer to the ground than gazelles do. When I approached, I discovered a whole bunch of jumping hare hopping around.

It was always a pleasure to detect one of the pretty aardwolves among the gazelles at night. Since I saw single aardwolves fairly often in the immediate vicinity of tommy herds, I suspected that they might be "interested" in gazelle fawns. With this ugly thought, however, I did great injustice to these harmless animals. They definitely feed only on termites and insects; at most they may sometimes take a bird's egg, a lizard, a mouse, or something of this size.

Not as harmless were the spotted hyenas. One night, when I was sleeping in the back of my car as usual, I was awakened by strange rocking movements of the vehicle. Close by I could hear the famous "laughing" and "giggling" of hyenas, uttered mainly when they are strug-

The eyes of jumping hares flash up in the beam of a searchlight.

gling over food. When I switched on the lights, several hyenas promptly fled from the car, but they soon returned and continued their activity. I feared that they were biting holes in the tires, so rushing out of my warm sleeping bag, I jumped out of the car into the pitch-black and very cool night with a panga (a long knife, similar to a machete) in my fist. Loudly shouting, I chased away the hyenas, and they disappeared in the dark. I was hardly back in the sleeping bag, however, when the rocking started again. This nice play was repeated until I eventually became tired of it. I let the hyenas be hyenas, remained in my sleeping bag, and fell asleep—more or less mildly rocked as in a cradle. In the morning, I discovered to my great relief that the hyenas had not bitten the tires. Instead, they had chewed the (switched-off) parking lights down to little relics! In the course of three years in Serengeti, I spent almost three hundred nights sleeping in my car in the field, and there were always hyenas around, sometimes only a very few, sometimes quite a number. Neither before nor after did they ever attack my car. I was never sure what came into those hyenas' "minds" that night.

When I once spent a night in the dense forests at the slopes of the Ngorongoro Crater, I heard again and again a strange cackling followed by almost ghostly piercing cries which apparently came from the tops of the trees. By chance, the next evening I had parked my car below a tall tree, when the "concert" started above me. It was still light enough that I could recognize the "singer" sitting on a branch: a nice old tree hyrax.

One night, I observed tommies in the Togoro plains until they bedded down about three o'clock in the morning. I knew that they now would rest for the next three hours. So I intended to doze awhile but postponed it because I was fascinated by the view of a topi bull standing like a sculpture of the noblest material on top of a termite mound, bathed in bright moonlight. While observing the gazelles, I several times had heard a soft noise which apparently came from beneath my car, but I had paid no further attention to it. Now, while I was staring at the beautiful picture in front of me, a small animal dashed out from under the front wheels of my car, ran a few yards, stopped, turned around, and barked at the vehicle. It was a bat-eared fox. I had unintentionally parked my car right above the exit of his den so that the little guy apparently had not dared to leave it for quite a while. Eventually, however, he took courage. Now he was loudly expressing his well-justified indignation.

When I was at home nights at Banagi, I liked to spend some time watching my "neighbors," a pair of dik-dik. They had their territory, which was about four times as large as an average tommy territory, around my house. During the day they rested somewhere hidden in the bushes and other high vegetation. In the evening, they came out of their cover in a nice "passing play." First the little female moved in a brisk walk over some distance, while the little male remained standing behind. Then she stopped and he moved ahead, passing her and continuing for a distance in front of her. Now he stopped, and she started walking again. She passed him while he was standing, moving ahead of him, and so on.

Not quite twenty yards away from my house, there was a single huge tree. Below it, the dik-dik had one of their regularly frequented dung piles. They usually arrived there right around nightfall. Comfortably sitting in an armchair, I could study their "marking ritual" at close range from the window of my house. Followed by the male, the female came up to the dung pile, crouched, and urinated and defecated there, placing her excrement on top of the already present dung. The buck stood waiting behind her until she was through. Then he sniffed at her urine and droppings, scratched in the dung with one foreleg, and placed his urine and, deeply crouching, his droppings on top of it, whereby he circled around his own axis several times. After this, he regularly marked

"My" dik-dik pair at Banagi.

the same long grass stem nearby with the secretion of his preorbital glands. In the course of time, a thick blackish-brown "pearl" of dried-up secretion stuck at the top of this stem. In contrast to Thomson's gazelle, dik-dik females also mark objects with their preorbital-gland secretion, although much less frequently than the males.

After some time, a fawn was born in the territory of "my" dik-dik pair, and its favorite spot to lie out was under my house. To better protect the "bungalow" from ants and termites, it was built on posts so that there was a free space between its bottom and the ground. The young dik-dik grew fast, and after six months he had reached three-quarters of the size of an adult. Little horns characterized him as a young male. At this time, the father began to "take offense" at the presence of his son. He still tolerated him in the territory, but he turned nasty when the young male tried to accompany his parents on their "circuits" in the evening and at night. Particularly when the young male approached

The dik-dik father rushes toward his (almost) grown son with his crest erect. The son lies down in submission.

the female, his mother, probably still entirely with "infantile intentions," the adult buck rushed toward him with his crest erect. (Dik-dik have a crest of long hair on the forehead which can be erected in excitement.) In response, the youngster lowered his head and neck and bent his legs, or even lay down completely with his head and neck stretched forward on the ground. This submissive behavior regularly stopped the adult buck's aggression. He stood and looked at his submissive son like a chicken at an earthworm, first with the one, then with the other eye, his crest went down, and he turned away. The son arose and/or assumed his "normal" attitude, respectively. Sooner or later he approached his mother again, and the whole procedure was repeated. On one brightly moonlit night which I devoted to the observation of "my" dik-diks, I counted fifty-two such encounters.

From the day on which I first observed the father "take offense," the son remained in the parental territory for about four weeks. After that, apparently the harassment became too much for him, and he left. However, he did not go far but remained in the "no-man's-land" between the territory of his parents and that of the next dik-dik pair. I often met him there later.

As can be inferred from the above, dik-dik live in pairs and in permanently maintained, relatively large territories. They are predominantly, though not exclusively, active at night. On dark nights they have several activity bouts between longer resting pauses. In bright moonlight, they may remain on their feet from the evening until about three o'clock in the morning. I always felt as if I had been transported into a fairyland when watching these charming dwarf antelopes in the magic glamor of an African moonlit night.

I am writing down these "nocturnal impressions" while sitting at my desk in my house in a northern forest area. Night fell half an hour ago. In the Serengeti plains, the gazelles have bedded down by now for their first nocturnal resting phase. When I close my eyes, I once again can see their eyes flashing up in the beam of the searchlight. The eyes of the gazelles are shining yellow-green. . . .

6 On Mothers and Their Young

Young antelopes are either "followers" or they are of the "lying-out" type. Followers stay continuously with their mothers and follow them, whereas lying-out young rest alone somewhere and come together with their mothers only for nursing. This is a very rough outline which badly needs completion, but as a first introduction it may do.

The fawns of Grant's and Thomson's gazelles are of the lying-out type. They are charming little creatures. In proportion to the skull, the nasal region of very young fawns appears shorter and more compact than at a later age. In my private vocabulary I called them "little box-noses." Also, for at maximum two weeks, the neonates are of a darker coat color than adult animals and the older fawns.

Normally, giving birth does not take a long time in gazelles. In the field, I became aware of the beginning of a birth when the amniotic sac protruded as a "white ball" from a female's vagina. From this occurrence to the completed expulsion, it usually takes about half an hour—sometimes somewhat longer, more often a little shorter. Soon after the expulsion, the fawn starts struggling to get to its feet, and after twenty to

A lying-out Grant's gazelle fawn.

thirty minutes, it usually succeeds and stands on its legs for the first time. The time of the first nursing depends on several circumstances, not least the mother's support. An experienced mother pushes her neonate with her nose in the direction of her udder. It is probably correct to say that a gazelle fawn commonly is nursed for the first time about fifteen minutes after its first standing. In rare cases, the neonate may suckle while it and the mother are still lying down. The mother regularly eats the afterbirth, which she may expel shortly after giving birth, but sometimes not until two or even more hours later.

A female gazelle close to giving birth separates from the herd, often simply by staying behind when the others move on. Thus, usually she is alone in this situation. Occasionally, however, other gazelles happen to pass along or to arrive there when the birth is underway. In such a case, the mother does not take any measures against them, nor does she move away. However, it is hardly an anthropomorphism to say that the company of the others is anything but welcome to her at this time. Quite frequently a few curious subadults are among the newcomers, and they take a keen interest in a female in labor. They sniff her genital region, mount her, and bother her otherwise, and when the young is born, they do not treat it gently.

While giving birth, the female alternately stands and lies down. In standing, she may lick the amniotic fluid from the grass. In lying, she typically rests on her flank with fore- and hindlegs stretched sideward away and her neck erect. Spasmodic stretching movements of her legs

A neonate tommy fawn struggles to its feet, is eventually pushed by the mother's nose in the direction of her udder, and is nursed by her.

and neck may occur during labor. Once the fawn's shoulder region is out, the rest of the expulsion goes easily. It usually is initiated when the female is still lying down, but it is completed while she is standing. Thus, although the mother assumes somewhat of a crouching posture at this moment (similar to when she is urinating), the fawn drops down to the ground, whereby the umbilical cord usually tears.

Immediately afterward, sometimes during the birth itself, the mother turns toward her fawn and licks it. In this way she removes amniotic fluid and tissues. Furthermore, her "tongue massage" may stimulate the neonate's blood circulation and, thus, may be helpful in its attempts to stand. Possibly, the mother also "labels" her young with her saliva so that she later can recognize it olfactorily and distinguish it from other fawns. Sometimes a mother's intense licking seems to thwart the neonate's first rising. On the other hand, once when a tommy fawn was slow to get up, the mother licked its forehead intensely and persistently until it rose as if pulled up by her tongue.

In young of the lying-out type, the first three to six hours after birth are often atypical, in that mother and offspring may remain together, rest together, and walk together, whereby they leave the place of the birth. At the latest after six hours—but sometimes considerably earlier—the first instance of lying out takes place. The most important criterion of lying-out behavior is that the young, completely on its own, turns away from its mother, moves ten to fifty yards away from her, and lies down. A gazelle mother watches her fawn during its walk away, and if it disappears from her sight in the grass or between bushes before it has bedded down, she follows a few steps until she can see it again. After the young has lain down, she often calls. When the young does not rise and come to her after this calling (if this happens, the whole procedure is repeated a little later), she turns and slowly moves away, grazing. She may easily stray fifty to a hundred yards, but sometimes much farther. If she beds down later, she rests far away from her fawn. In contrast to some forest animals, the mother's contact with her lying-out young is not interrupted completely in gazelles in the open plains. A gazelle mother keeps her fawn's resting site and its vicinity under watch. She becomes alert and approaches as soon as something seems to be going on there—be it only that the fawn rises after a while in order to immediately lie down again in another position.

Usually the lying-out procedure works just perfectly, even on the first occurrence. Only once did I witness a case among the gazelles in Serengeti in which this was not true. It involved a tommy female with a fawn no more than a few hours old. Besides its small size and dark color, this fawn was still somewhat shaky on its feet, and it several times tried to suckle in the wrong place, between the mother's forelegs. If such a mistake happens at all with a fawn only a little older, it happens once

Above: The fawn turns away from the (grazing) mother, moves away from her, . . . *Below:* . . . and beds down for lying out, while the mother watches it (Grant's gazelle).

now and then but not several times successively. When this fawn eventually found the mother's udder, she nursed it and licked it, then stood, apparently waiting for it to turn, move away, and lie down. However, this fawn did not do so but lay down after a while right at the mother's forelegs. When she moved off, it stood up and followed her, and when she stopped walking, it lay down again, this time at her hindlegs. This "nice play" was repeated several times. Finally, the mother, followed by her fawn, walked to a tall tree, the only one in the surroundings, and remained there standing. After a few minutes, the fawn took two or three steps toward this tree, lowered its head, and lay down in the space between two roots branching out above the ground. The mother stood there until her fawn had fallen asleep. Then she moved away.

Usually, a tommy mother keeps the lying-out place of her fawn under watch, although at a distance. She becomes alert as soon as something seems to be going on there—be it only that the fawn rises and beds down again (in another position). The mother approaches (grazing in this case), looks to see whether everything is all right and, seeing that it is, moves away again.

One reason this event was so interesting to me was that years ago, I had made a comparable observation with Indian nilgai antelope in a zoological garden. As is quite common in nilgai, a female had given birth to twins. These two little calves stuck with their mother and lay down at her legs again and again, always the one calf at her forelegs and the other at her hindlegs, so that this mother also had a lot of trouble getting her calves into the "right" lying-out situation. In this case, however, the young had been born in a narrow box in the stall, and they had remained there together with their mother for two days because of bad weather. Therefore, when they came into the outdoor enclosure on the third day and showed this "faulty" behavior, I presumed that an "imprinting failure" had taken place during the unnatural situation in the stall and that this failure was due to the captive conditions. Now I had learned by the example of the tommy fawn that similar things may also occur in the wild, although probably only exceptionally.

Obviously, one favoring condition for lying out is that the young is being nursed. Thus, the mother can induce or increase the fawn's readiness to lie out by nursing it. As long as the fawn is with her and on its feet, she also can lead it to an area suitable for lying out, or she can bring it up again by pushing it when it has bedded down in a place not to her liking. Otherwise, however, she does not have any influence on the process. The turning and walking away from the mother as well as the choice of the specific resting place are independent activities of the young. This fact can hardly be overemphasized, because it is the deciding difference in the behavior of followers, in which it also may happen occasionally that a young rests separated from its mother (see below).

When searching for a suitable place for lying out, a fawn often responds to certain environmental stimuli. The two most important of these are "something vertical" and "something dish-shaped," i.e., a small depression in the ground. "Something vertical" may be a tree, a bush, high grass, a rock, a big stone, a termite mound, or the like. "Something dish-shaped" might be a ditch, a hole in the ground, the space between roots, or an erosion step; to a little gazelle fawn, even the bare ground between two bunches of short grass may be sufficient. In the optimal case, the two "stimuli" occur together, for instance, when there is a tree with roots sticking up above the ground at the base of the trunk. In this case, the fawn will predictably bed down right at the trunk (something vertical), nestling between its roots (a dish-shaped excavation). However, in principle, one of the two characteristics, something vertical *or* something dish-shaped, is enough. When, after some searching, a gazelle fawn has found neither something vertical nor something dish-shaped, as may sometimes happen in the plains, it will lie down right in the open. Obviously, the most important point to the fawn is to move away from its mother and then lie down. Being visually hidden while lying out appears, from the animal's point of view, to be more a secondary matter, although it may be quite useful and beneficial to the fawn's survival. (Thus I think it is inaccurate when certain people refer to lying-out young as "hiders" and/or to lying-out behavior as "hiding.")

"Lying out" is not the same as "shock-lying." The latter occurs predominantly in very young animals, including at least some followers. It is induced by the sudden appearance and/or approach of a strange object, or sometimes by a sudden loud noise. The lying young stretches its head and neck forward or oblique-forward and either presses them on the ground or keeps them freely but stiffly a very few inches above the ground. It is in a state of rigidity which, during the first hours or days of life, is kept even when, for instance, a human approaches the young to touch and pick it up. Somewhat older young jump up eventually and run away when the suspicious object has approached them within two to three yards. In tommy fawns, this change of behavior occurs at the latest twenty-four hours after birth. In other species it may take a few days.

In shock-lying, standing or walking young throw themselves down abruptly, which sometimes may result in rather strange lying postures. Lying-out young may change from lying out to shock-lying. Since this happens regularly when a human approaches them, some people think that these young were always resting with their heads and necks stretched forward. Of course, this posture can be assumed in lying out, since it is one of the normal resting postures in these animals. However, when lying out undisturbed, at least as often a fawn bends its neck sideward-backward and puts its head on its thigh or beside its hindleg, or it may keep the neck erect while dozing.

In some species, shock-lying may also occur in adult animals. Famous for it are the duikers, who got their name from this behavior. (*Duiker* is a word from Afrikaans meaning "diver.") It is also found in some other small antelopes, such as the little steenbok. Occasionally, however, it occurs in larger or even very big animals. Once when I was walking through a dense forest area, I suddenly encountered a fully grown eland antelope bull at a distance of a very few yards. Obviously, the bull was as startled as I was. My surprise became even greater when this gigantic fellow dropped down to the ground right in front of me and remained there lying flat.

When a lying-out young has rested alone for several hours, its mother calls it for the next nursing. A gazelle mother approaches slowly and very "casually," grazing in a zigzag course. She does not come up to her fawn but stops, stands, and calls it at a distance of ten to thirty yards. I did not actually hear this call from Grant's and tommy females in the wild. I knew it from dorcas gazelle in a zoological garden. There, I could accompany a completely tame female, side by side, as she approached her lying-out fawn, and I could clearly hear her call uttered by nose. (Probably this call—as well as most of the other vocalizations—is somewhat louder in dorcas gazelle than in Thomson's gazelle.) In the wild, as I said, I never came so close to a gazelle mother in this situation that I could hear her calling; however, I could see it. Watching such females by telescope, I could recognize how the skin above their nostrils went up for a moment, which always happens when gazelles utter vocalizations by nose. Additional evidence was provided by the behavior of the fawns in some cases. Occasionally a lying-out fawn would arise when the mother had approached to and now stood at the described distance, but the fawn looked in a different direction and did not approach her. Only when the mother made a move, for instance, a few further steps in the fawn's direction, did it look at her and come running. Thus it appeared that the fawn's first reaction (rising) was merely due to an acoustical perception.

When the fawn, sometimes trotting, sometimes at a gallop, arrives at the mother, she briefly sniffs at its nose and then more minutely at its anal region. Again I know from the observation of closely related species in zoological gardens that the mother can recognize in this way whether the little one is her own or a strange fawn. When a strange fawn has come up to her—which apparently happens less frequently in the wild than in captivity, where the spatial limitations facilitate such mistakes when several young are present—she pushes it away immediately after this olfactory inspection. While the fawn is suckling, the mother licks its anal region. With a young male, she will also reach with her nose between his hindlegs under his belly and lick his penis. In female fawns, she cleans anal and genital region in one stroke. Even when observing from a distance, it therefore is possible to determine the sex of the young from the mother's behavior. This maternal licking stimulates the young to urinate and to defecate. Thus, its excrement is deposited in a different place from that where it will rest later. Furthermore, the mother removes particles of urine and feces from its coat in this way. During the very first days of the fawn's life, she even drinks its urine and eats its feces. When later the fawn is lying out again, it not only visually "disappears" from the scene but is also as odorless as possible.

As mentioned above, I had observed the mother's calling for her fawn in several gazelle species in zoological gardens, and I had also taken pictures of this procedure there. I never dreamed that I would succeed in taking photos of such an intimate scene of the animals' life in the wild. But one day I had this luck. I had parked my car as usual in an area well populated by gazelles and was watching a herd to the right of the car. To the left, there were no animals for the time being. After about two hours, I happened to glance to the left, and I saw a tommy female approaching slowly, grazing, and looking around from time to time. Her behavior looked so much to me like the approach toward a lying-out fawn that I prepared myself to take pictures, although I did not see a fawn. Suddenly, a little head popped up from the short grass not quite twenty yards away from my car. I had been close to this fawn the whole time, but I had not become aware of it. When it now arose, looked around, and ran toward its mother, I was able to take pictures of the whole story.

After nursing, mother and young remain together for half an hour to an hour. During this time, a gazelle fawn often plays around the grazing mother in stotting gaits and rushes. I always admired the speed which such a little chap can reach in its running play. When the mother moves ahead, the fawn follows her. Thus, the following response is not lacking in young of the lying-out type; however, in comparison to followers, its occurrence is not only restricted to a relatively short time but often also gives the impression of being not quite as perfect. A young follower walks predominantly behind its mother or beside her next to her hindlegs. At most, it comes forward along the mother's body so that its head is even with her shoulder. By contrast, a gazelle fawn may sometimes pass its mother and walk or run in front of her for a little while. Particularly when mother and fawn are moving at an accelerated speed, they frequently pass each other alternately all the time. Sooner or later, however, a gazelle fawn turns and walks away from its mother, searching for a place to rest until the next nursing.

With increasing age of the fawn, the times it spends with its mother become longer and longer and its lying-out periods correspondingly shorter until this behavior comes completely to an end. In gazelle fawns, this occurs at an age of about two months. At this age, the young has already begun to take solid food, although it may still suckle—if the mother still has milk. A gazelle fawn may now solicit milk from its mother pretty vehemently. Standing or walking beside her, it may push with its forehead against her belly, or it may walk behind her and kick with its foreleg against her hindlegs—very much like a courting male—

A TOMMY MOTHER CALLS HER LYING-OUT FAWN FOR NURSING

Above left: The mother approaches her lying-out fawn (whose head is visible in the left corner of the photo). *Above right:* She calls at a distance. The fawn rises but has not yet looked in the mother's direction. *Center left:* The mother approaches somewhat closer. Now the fawn sees her. *Center right:* The fawn gallops toward the mother. *Botton left:* Upon its arrival, the mother briefly sniffs at the fawn. *Bottom right:* She nurses and licks it.

or it may mount her. Particularly in Thomson's gazelle, a somewhat older fawn can bother its mother so much that she will flee at a gallop, pursued by her fawn.

Nevertheless, the bond between mother and young remains, probably even longer in the gazelle species under discussion than in certain other ungulates because the fawns of Grant's and Thomson's gazelles do not form "kindergartens," and thus the bond with the mother is not weakened or even replaced by contacts with mates of the same age. Of course, gazelle fawns also have sparring matches; sometimes one of them will climb a termite mound and "defend" it against several others, and their running plays can be "contagious" so that several fawns are often running at the same time. In this playful running, however, each gazelle fawn takes and keeps its own course. Thus, for instance, playful chasing in which one young runs after another, as is frequent, for example, in topi, is lacking in these gazelles. Occasionally a play "bout" may last up to half an hour, but commonly it is over after five to ten minutes. Then even an already half-grown gazelle fawn returns to its mother, and—after the lying-out period is over—it spends most of its time near her within the herd. In grazing, it frequently stands right beside the head of the grazing mother, and it may learn to take her preferred food plants in this situation.

Provided that mother and young are not separated from each other by external forces and that one of them does not suffer a premature death, a young female gazelle may remain with her mother well past her first year, possibly even considerably longer; of course, the bond between them becomes somewhat looser in the course of time. Maturing male fawns would probably also stay with their mothers for quite a while, if they were not separated from them by the activities of the territorial bucks. While the owner of a territory does not take much notice of half-grown fawns up to an age of about five months—and even less of younger fawns—he takes adolescent males beyond this age more and more seriously. Thus, when in Thomson's gazelle a female herd in which there is a mother with a son of the corresponding age enters a territory, the territorial buck chases the youngster out of his territory. In the beginning, the mother may follow her son in such a case and wait outside the territory until the other females leave it so that she and her son can rejoin the herd. However, the older the young male grows, the more promptly and frequently he is chased away when entering a territory; one day the mother will not follow him anymore, and he may now join an all-male group. As I observed several times, a tommy mother may occasionally even accompany her son into an all-male group. Here, however, being the only female among many males, she is immediately and very intensely "courted" by up to six or seven adult bucks simultaneously. This is too much for her, and she will flee back to the area where the territories are and where the bachelors do not dare follow her. Her son may follow her, only to be promptly chased by a territorial buck again. So the "play" may go to and fro for a while, until eventually the mother remains with the other females, and the young male stays with the bachelors, and thus the separation is completed.

When a gazelle mother approaches her lying-out fawn for nursing, as described above, she does so quite casually and slowly while grazing. Thus, she has time enough to look around, and it may be said that she takes some care before calling her fawn. However, this is nothing compared to the precaution which I experienced on the part of an oryx mother. Admittedly, I do not particularly like to tell this story, since I behaved rather clumsily and not particularly intelligently in it, but it is very instructive with respect to the animals. Oryx antelope are not permanent residents of the Serengeti National Park, but they pass through it from time to time. All the migratory oryx herds I observed consisted of adult bulls and cows. Subadult or even younger animals were not with them.

Once I was watching a herd of twenty-two oryx antelopes. In the afternoon they were dispersed over an unusually wide range, walking and grazing. I was with the head of the group, and only now and then did I look back with my binoculars to the last stragglers just still recognizable in the wavering air. Once I had the impression that a very small animal was far out there with the oryx, but I did not pay special attention to it. The situation remained unchanged until nightfall. It was a dark night with a cloudy sky and without any moonlight. Lions were roaring far away in the gallery forest of Ngare nanyuki. There was still enough light for me to recognize that the number of animals around my car had increased. Obviously, the stragglers had arrived. Otherwise I could hardly see anything, and thus I eventually went to sleep.

When I woke up the next morning, the herd had moved to the horizon, and it soon disappeared from my eyes completely. Only one cow remained behind, grazing alone about eighty yards away from my car. At this sight, I remembered the little animal I had seen from such a far distance yesterday, and it came to my mind that this oryx cow might possibly have given birth to a calf which was now lying out somewhere in the surroundings. Therefore, I decided not immediately to follow the herd but to stay with this cow for a while.

About 9:30 A.M., she stopped grazing and slowly paced away—however, not in the direction in which the herd had gone but in the direction from which it had come the day before. I followed her at a distance of about one hundred yards, and she tolerated it without being obviously disturbed. I now believed she was going to her lying-out calf and would lead me to it. The oryx female moved slowly, pausing numerous times to look around attentively. Once, she even bedded down and rested for a little while. By 11 A.M., she had walked about two miles and had arrived at the rim of a rising slope relatively elevated in this open plains area. There the oryx cow stood for a long while, carefully watching out to the far-out gallery forest of Ngare nanyuki, where lions had been roaring the evening before. Suddenly she turned and, partly trotting, partly at a gallop, hurried all the way back to where she had been grazing in the morning. Upon arriving there, she stopped, stood still—and about twenty yards in front of her, a very small calf arose from the short grass and ran toward her. The mother sniffed at her calf, licked and nursed it. Although I had been near its lying-out place for hours, I had not seen anything of it. The mother's long march had led not to the calf but away from it. It had been a very intense, precautionary, and purposeful reconnaissance.

I had followed the oryx cow on her way back as well, and now I was standing with my car about sixty yards away from her. I wanted to take a picture of this mother nursing her calf. As usual, I had a sandbag with me, which I used to put on the lower frame of the car's opened side window as a safe and firm support for the camera and the telephoto lens. However, since the calf had already suckled for several minutes, I was afraid that I would be too late if I went through the whole procedure with the sandbag. So I did it without it and probably somewhat too hastily. In any case, the camera with the long telephoto lens slipped out of my hands and dropped outside the car. The oryx cow, who had taken the presence and the moves of my car so well and calmly up to now, was panicked and fled at a full gallop, followed by the calf at her heels. After a mile or so, she stopped, looked back, and then continued the flight with her young. In persistent gallop the two reached the horizon and disappeared from my sight. It did not help much that I loudly called myself bad names. I did not dare follow the mother and her calf because I thought it very possible that this oryx cow now had become "allergic" to my car and would run again upon seeing it. This, however, I did not wish to do to the calf. It had been admirable enough how this little chap had run after his mother and kept up with her even at the fastest gallop. I only hoped that he had not suffered any heart trouble from this persistent high-speed performance at an age of hardly twenty-four hours.

Wildebeest calves following their mothers.

Good luck within bad luck: my camera and lens had not been damaged by falling from the car window.

Although the young of most game animals in Serengeti—such as Thomson's gazelle, Grant's gazelle, impala, steenbok, dik-dik, kongoni, reedbuck, waterbuck, bushbuck, and others—belong to the lying-out type, there are also some quite pronounced followers, such as zebra, wildebeest, and topi. The mother-offspring relationships of wildebeest deviate on several points from those of the majority of East African ungulates. For instance, wildebeest have a relatively limited calving season. Most of their calves are born within a period of four to six weeks in January/February in the Serengeti area. Also, the highly pregnant wildebeest cows do not separate from the others but give birth within the large herds. Furthermore, a wildebeest calf usually stands firmly on its feet within five to ten minutes after birth. Finally, wildebeest mothers do not commonly eat the afterbirth; they probably would like to do so, but in the very best case, they can take only a few bites, and then numerous vultures drop down from the sky and take care of the rest.

When its mother moves, a young follower stays right behind or beside her. When she rests, her calf lies next to her. Only when she is grazing may a spatial separation between the two occur. A young calf does not take solid food, at least not to a noteworthy extent, during the first days and weeks of its life. On the other hand, it needs more rest and sleep than the adult animals, which spend much time in feeding. So it happens almost automatically that a little follower beds down while its mother is on her feet and grazing. In this situation, the mother sometimes may move away, depending on circumstances up to distances approximating those between mothers and young of the lying-out type. Thus, a temporary separation between mother and offspring per se is not reliable proof that the young belongs to the lying-out type—a point which, in my opinion, is not too hard to understand but which apparently cannot make its way into the brains of certain otherwise very intelligent persons. The decisive difference is that the active process of turning and moving away from the mother is lacking in followers, and thus their possible separation from their mothers is brought about in a different way—it may even be said, in the opposite way—than in the young of the lying-out type. Furthermore, when the mother of a follower has grazed at a distance long enough and her calf is still resting, she returns to her young and beds down near it, whereas the mother of a lying-out young does not do so (except during the very first hours after birth, as described above).

Despite the well-developed following response of wildebeest calves, it happens in the gigantic herds on special occasions, for example, when crossing a ford, that some of the calves lose contact with their mothers. The chances that they may find each other again are slim in the crowd, and since a wildebeest cow normally nurses only her own young, such calves are doomed to destruction. Usually they don't have to "wait" until they die by starvation; the following night, the numerous hyenas, lions, etc. will "take care of them" in Serengeti.

In principle, zebra behave like the horned ungulates of the follower type. However, there is at least one very typical difference which struck me every time I saw it. Resting completely on the flank with all four legs stretched sideward and one side of the neck as well as the one cheek on the ground, is extremely exceptional in all the antelopes. Studying quite a number of such species in the wild and in captivity, I have seen it perhaps four or five times in thirty years. In zebra, however, and particularly in their young, it is not rare. Furthermore, I have the impression that adult zebra do not lie down as readily as most of the ruminants do, but they apparently doze and even sleep standing up more often than the even-toed ungulates. Taking all this together, a "picture" frequently seen emerges which I do not know from any antelope species: During a resting phase, the young zebra lies flat and completely on its flank, and the mother stands next to it with her face toward it.

If a follower's mother moves away from her resting young, this usually happens step by step in grazing. Relatively infrequently, a follower's mother will leave her young intentionally and/or at a straight walk. I

A zebra mare guards her young, which is resting completely on its flank.

once had a quite amusing experience. Years ago, I had raised a young blesbok (a South African species related to the East African topi) in the Zurich zoo in Switzerland. It was a female calf, and I named her "Mousy." She stuck so firmly to me as her "mother" that I literally could not leave her alone for one minute—except when she was dozing or, even better, asleep. I made use of this as often as possible, since at least sometimes a man has things to do for which a blesbok calf is not quite the ideal companion. For instance, I regularly went with Mousy to a large meadow at noon, where she first performed marvelous running and jumping games, and then lay down to rest for about half an hour. Then I could leave her, but I had to do it very carefully because, as soon as she became aware of my departure, she came dashing after me. The "critical distance" was the first thirty yards. These I had to cover inaudibly, in a straight line (like all the ungulates, Mousy could visually discriminate sideward movements much better than movements leading straight toward or straight away from her), and, above all, very, very slowly. When I had made it in this way up to a distance of about thirty yards without evoking her following reaction, I could move freely. At this moment, I always felt such relief that I now ran away with long jumps.

One day in Serengeti, I observed a topi female followed by a calf of at most two or three days of age. They walked a distance, then the mother stopped, and the calf lay down beside her. The mother stood still for a little while, but eventually she started slowly moving again. The calf arose and followed her. She stopped, the calf lay down—and the whole sequence was repeated several times. Remembering my "life with Mousy," I could not suppress the feeling that this topi cow would have liked to leave her calf alone for a while. At the same moment, I remembered the criticism of a "dear colleague" on a comparable occasion. He had said: "You are not an antelope. Thus, you cannot know what they like or dislike." Modifying a saying of Chuang-tse, I had answered: "And you are not me. Thus, you cannot know whether I may know it." In the present case, however, I felt obliged to admit that, on the basis of such an "analogical conclusion," I might be seeing the situation of this topi mother in a too subjective way, and that I probably was anthropomorphizing her behavior. Meanwhile, the two had approached within fifty yards of my car. Again the mother stopped walking, and her calf lay down. This time, the topi cow remained standing longer than ever before. With the binoculars at my eyes, I could see the little calf close its eyes, and its head sank deeper until its snout vertically touched the ground—a posture frequently assumed by topi and their relatives when dozing or sleeping. Eventually, the mother began to walk—step by step, in slow motion, and in a straight line away from the calf, precisely the same way as I once had exercised it with Mousy. When the mother, having covered about thirty yards in slow motion, suddenly took off at a gallop, I could not help bursting out in laughter. The analogy was just too perfect. My first, certainly very subjective impression had obviously been correct, and I found—by the way, not for the first or the last time—that the parallels between human behavior and animal behavior can be more numerous than even an experienced observer may sometimes readily admit, and that such "analogical deductions" can be quite helpful in understanding the animals' behavior.

Something which is lacking in Grant's and Thomson's gazelles but is not unusual in some other species is "kindergartens" containing young of about the same age. In species whose young are of the lying-out type, such as impala, the somewhat older fawns form such groups of juveniles within the female herds. In followers, such as topi, the calves in a "kindergarten" can be very young. A few days after giving birth (like wildebeest, topi have a limited calving season), three, five, or even more topi cows may form a mothers' group. When such a group is resting, the calves lie in the center with their mothers around them, more or less in ring formation. Later, when the cows rise, start grazing, and slowly move away in doing so, the calves remain behind resting in a group. By

Often several topi calves rest together in a "kindergarten" while their mothers move away in grazing. Sometimes one of the cows remains with the calves as a "governess."

no means always but sometimes, one of the cows remains standing with the calves while the other mothers move farther away. That one female may take the role of a "governess" in this way has been claimed from time to time to be true in several game species, and it has been denied about as often. In topi, I have observed it several times and beyond any doubt. When the grazing mothers become mildly disturbed—for instance, by a slowly approaching car—they return to the "kindergarten." Each of them calls her calf, quite frequently she nurses it briefly ("disturbance nursing"), and then she walks away followed by her calf. This last scene reveals once more that the resting behavior of topi calves is not lying-out behavior because, except on very special and, on the whole, rare occasions, the mother of a young of the lying-out type does not approach, call her offspring, and make it rise when something disturbing and possibly even dangerous is around.

From the View of the Preyed-Upon

When talking about lions in a previous chapter, I mentioned that some people may form a definite idea about something which has little to do with reality, but they firmly adhere to it even when they have contradictory evidence right in front of their eyes. Unfortunately, this psychological "mechanism" is sometimes effective in certain scientists as well. In this case the "ideas" usually are their so-called "hypotheses," and from all the available facts, they carefully pick everything which fits these hypotheses but just as carefully ignore and "forget" everything that does not fit. In this regard, it would also be correct to include "survival of the fittest." I found this "mechanism" to be particularly prevalent when the discussion turns to predator-prey relationships—at least as they occur among the game animals in Serengeti. Not only laymen but sometimes even professional and otherwise highly distinguished zoologists express views which are wrong or at least extremely debatable in the light of facts. I always read such publications or listen to such opinions in part with amusement, in part with regret, and in part with downright furor because at least some of these authors truly should know better.

To begin with, there is the readily accepted and frequently repeated fairytale that the horns of antelopes and other horned ruminants are weapons used primarily in self-defense against predators, and that they phylogenetically evolved for this purpose. When you ask the conveyers of such "wisdom" how many fights between horned ungulates and predators they have witnessed (with the exception of maternal defense of young, which is by no means restricted to the possession of horns, and about which we will speak later in detail), the overwhelming majority of them, as far as they are honest, have to confess that they have never seen even a single case of this kind, even when they have spent months and years in the field. Those few who have witnessed an antelope fighting a predator usually can cite only one case, or very exceptionally two or three. On the other hand, anyone who has had the opportunity to observe a gregarious antelope species for even one week will probably have recorded at least ten fights among conspecifics. With some luck, he may observe two or three times that many in one day. In short, the extremely small number of fights between horned ungulates and predators is out of all proportion to the numerous fights among conspecifics. On this basis, a man could easily deduce that the horns of these animals are not used primarily in interspecific* fighting but in combat with opponents of their own species, and that it is absolutely possible that they evolved in this context!

Furthermore, of the more than one hundred species of horned ungulates (Bovidae), roughly speaking, in only one-third of them do the females have horns even approximately equivalent to those of the males (in the East African game animals, for instance, wildebeest, topi, eland antelope, etc.). In another third, the females have considerably smaller and thinner horns than the males (for example, in Grant's and Thomson's gazelle), and in the last third, the females have no horns at all (as in waterbuck, reedbuck, bushbuck, impala, etc.). If the horns played any important part in the defense against predators, this "under-privileging" of females would be absolutely incomprehensible, since the females as well as the males are attacked by predators, and the females have to defend not only themselves but also their young. We might presume that these not too complicated considerations would deal a death blow to the hypothesis about horns being primarily weapons for use

*"Interspecific" behavior denotes an interaction between animals of diferent species; "intraspecific" behavior refers to an interaction among conspecifics (animals of the same species).

against predators. However, adherents of this hypothesis are well prepared for such a blow and parry it with a "logical" argument: The females do not need horns, or at least not very strong ones, because they and their young are defended against predators by the well-armed males. And—how beautifully things fit together!—the many fights among conspecific males confirm this theory. These are necessary training for the defense against predators. Moreover, these intraspecific fights result in a selection among the males, and they actually breed for well-armed, strong, and brave defenders of the herds and the families, optimally fit to protect the females and their even more vulnerable young against attacks from predators. Anyone who wishes to learn more about such (to put it mildly) questionable hypotheses can consult, for instance, the corresponding sections in Konrad Lorenz's famous book *On Aggression.* (I mention here the name of Konrad Lorenz not because I wish to diminish the merits of this outstanding man but because I wish to demonstrate just by his example that, when it comes to predator-prey relationships, even the mind of a Nobel Prize winner is not safe from the "ideas" and psychological "mechanisms" characterized in the beginning of this chapter.)

Let us leave the muddy swamp of such speculations and return to the solid, dry ground of the East African plains. Let us meet gazelles and other antelopes there, and see what they have to say about these things.

In the overwhelming majority of cases and species—from dik-dik up to wildebeest and topi—these animals, including the "well-armed" and strong males, are caught and killed by predators without even trying to defend themselves. At maximum, once caught, such an animal may perform a few feeble butting movements in the direction of the enemy, which usually do not even touch him. Only in about ten percent of the many species of horned ungulates may some kind of self-defense occur which goes beyond that, almost always after the animal has first tried to escape by fleeing. Worldwide, only one bovid species shows such a defense more or less regularly. This is the Arctic musk ox, whose chances in defense against wolves are even quite good, according to reports by observers. Among the African horned ungulates, only in the African buffalo and in roan, sable, and oryx antelopes has any real effort at self-defense been observed in some cases. However, even in these few species, the chances of success are moderate. I know of no horned ungulate that possesses a fighting technique particularly adapted to defense against predators. On the contrary, if they fight predators at all, they appear to use pretty much the same techniques as in encounters with conspecific rivals. In fights between conspecifics, the attacker's horns hit, push, butt, blow, press, etc., against the opponent's horns in most cases, whereas in a fight with a predator who has no horns, the attacking buck or bull will directly hit his enemy's body with his blows, provided that his horns will reach. This, however, seems to be pretty infrequent. Unlike a conspecific, a predator does not remain standing in order to catch and return the attacker's horn blow; he dodges and may jump around the prospective prey to attack from the flank or the rear. A horned ungulate is poorly adapted to such fighting. Thus, in most cases, he will be killed despite his defense. Only now and then may he succeed in intimidating and/or exhausting the predator to the point that the latter backs off, and even rarer are cases in which he may seriously wound or even kill the enemy. Thus, when, for instance, an oryx antelope gores two or three out of an attacking pack of wild dogs before being killed and eaten by the others, this is probably the most such an animal can accomplish.

One may speculate whether horned ungulates might be more successful in fighting predators if they were defending themselves mutually and in groups. As a somewhat regular behavior, however, group defense is again known only in the musk ox. In African game animals, the Cape buffalo is said to do it occasionally. I myself once observed a small herd of female eland antelope join forces to attack a cheetah. Thus, very occasionally such things may happen. On the whole, however, group defense against predators is even more exceptional in these animals than individual self-defense.

Considering how weak the tendency for self-defense is in the bucks and bulls under discussion, one can imagine what their defense of females and young looks like. Often this is out of the question right from the beginning because in many species, the males form all-male groups which live more or less separately from the females throughout long parts of the year. Then, when a predator sets upon a female or young, there frequently is no male around. But when males and females are together, for instance in a mixed herd, and this herd is attacked by a predator, it is "every buck and bull for himself," and he runs as fast as he can. He does not "think" about defending the females and/or young. Also, such defense by fathers is lacking in gregarious antelopes simply because they do not form families with both parents and young. In small species that form pairs, such as dik-dik, a father-mother-offspring family may exist temporarily. However, even in such a family, the male hardly takes notice of the young. Only when it is a maturing young male does the father begin to take an interest in the son, which means that he treats him as a prospective rival and eventually makes him leave the parental territory. Also, defense of a young by a male that is not its father simply is not part of the behavior of any of the antelope species known to me. By way of a particularly convincing example, I will describe a scene from the life of Thomson's gazelle in some detail. Since several, basically independent circumstances must occur together, it is not a very frequent occurrence, but neither, apparently, is it extremely rare. I witnessed it twelve times in the course of three years.

As extensively described, some tommy bucks become territorial, and the female herds visit them temporarily. When a group of females has entered a male's territory, it may happen that a mother of a small fawn is among them, and that this fawn is lying out somewhere in the surroundings but outside the male's territory. Furthermore, it may happen that roaming jackals detect this fawn. It flees, and the jackals chase after it. The mother comes running as fast as she can to defend her fawn—sometimes with, sometimes without success. At this moment, the territorial buck may follow the mother at his fastest gallop—however, not in order to assist her in defending her young, but in order to pass her, block her path, and prevent her from leaving his territory. The vital scene "fawn in danger" is not present in a tommy buck's behavioral inventory, but the vital scene "female leaves territory" is present, and he acts correspondingly.

Due to the "misunderstanding" involved, this example is a particularly crass case. However, in principle, the same is true for most (all?) of the horned ungulates. In the subjective world of these bucks and bulls, young have no meaning. Possibly, or even probably, they recognize the young as being conspecific animals, at least when they are beyond the neonate age. But that is it, and there is no special action or reaction with respect to the young in the behavioral inventories of these males. Correspondingly, I have never seen a case of a male's defense of young in any antelope species, and my colleagues and the game wardens at six African national parks never told me that they had witnessed such an event. There are a few reports of this kind in the literature; however, they are anything but convincing, and in some of them it is possible to point out the mistake in the observation. For example, in his otherwise pretty good publication *A Game Ranger's Note Book,* A. Percival first rightly states that tommy females defend their fawns against jackals, but then he tells us that sometimes the father of the fawn follows the mother at a gallop in order to assist her in the defense of their offspring.

Obviously, Percival has made similar observations to mine. However, he has confused things which he actually saw with things he assumed. He has seen an adult tommy buck—and assumed this was the fawn's father. With a gestation period of five to six months, and taking the migrations of the tommies into account, it is unlikely that a tommy mother with a fawn is still in or near the territory of the male who once serviced her. When there is now a buck around, the probability is almost certain that this is not the father. Furthermore, Percival has seen jackals run after a fawn, a tommy female run after the jackals, and a buck run after the female. He assumed the buck was doing so to support the mother in defense of her fawn. And when somebody has produced an interpretation which is so logical, it no longer matters that he has never seen the buck make any attempt to attack the jackals, and, of course, he takes no notice of the buck's attempts to stop the female!

In contrast to the "well-armed" males, female antelopes sometimes defend their young against predators, or may at least try to do so. In wildebeest, this often constitutes a symbolic "demonstration of good will." For instance, a single hyena may stroll along a group of wildebeest cows with small calves at a distance of about sixty yards, as may happen without any hunt ensuing. However, this hyena suddenly changes course toward the wildebeest and runs toward them at full speed. The wildebeest cows, so far only having watched out in the hyena's direction, now flee at a gallop, each followed by her calf. The hyena catches up with a calf running behind its mother and grasps it. The mother turns around and rushes with lowered head toward the hyena, but without reaching it. She then turns and continues the flight following the others, while the hyena kills her calf.

Although much smaller than wildebeest, the females of Grant's and Thomson's gazelles defend their fawns more energetically and persistently. Of course, they do not defend them against predators that are hopelessly superior to them, such as lions, leopards, cheetahs, and wild dogs. When her fawn is chased and caught by one of these enemies, a gazelle mother will watch excitedly at a distance of about one hundred yards, but she will not intervene. However, she will defend her fawn against smaller predators who are not much of a threat to adult gazelles and do not commonly hunt them. In Serengeti, these are predominantly the two jackals, the golden jackal and the black-backed jackal, who singly or in pairs search an area for lying-out gazelle fawns by moving in a zigzag course. Upon their approach, a fawn will press its body, neck, and head to the ground in shock-lying. Unless the jackals come directly upon such a fawn on their course, they usually will not become aware of it. However, when they come directly toward a fawn, it will jump up at the last moment and flee, often with loud roaring distress cries from its open mouth, and the jackals will start to chase it. The mother comes

A jackal has detected a lying-out tommy fawn and is chasing the fleeing youngster. The fawn's mother comes dashing in order to defend her young.

dashing, runs between her fleeing fawn and the pursuing jackals, and attacks them vehemently and persistently with blows of her thin, small, and—in tommy females—often broken or deformed horns. Occasionally a jackal will be hit hard enough that it rolls head over heels on the ground, but it is rarely injured otherwise. Most commonly, the jackal is able to turn aside skillfully and in time to evade the blow. Thus, the gazelle mother usually does not injure or kill the jackals, but she may succeed in diverting and exhausting them so that they eventually give up the hunt for this fawn. In a few such cases, I saw the jackals turn toward the mother and try to attack her, but always without success.

In gazelles generally, only one female, according to all circumstantial evidence the mother, defends a fawn, even when many females are around. Occasionally a second female may participate in the defense. It is not proven, but I think it possible that this may be a grown-up daughter of the fawn's mother or the mother of the fawn's mother. Extremely exceptionally, more than two females are involved. I witnessed only two events of this kind, and in both of them I had the suspicion that it might have been an "error" on the part of most of the females. Although Grant's and Thomson's gazelles reproduce all year round, there are seasonal peaks and declines in fawning. The observed events coincided with peaks of fawning, and thus several very young fawns were lying out in the vicinity of the herds. When the jackals had detected and begun chasing one of them, several mothers may have "presumed" that their own young were in danger. By the way, the hunted fawns were killed in both cases.

When only one jackal is after a gazelle fawn and this fawn is at least one or two days of age, the chances that the mother may defend it successfully are quite good. The chances decline when, as happens frequently, a pair of jackals are hunting, since while the gazelle mother is occupied with one of them, the other may find an opportunity to catch and kill the fawn. Nevertheless, successful maternal defense is still a definite possibility in such cases. The chances are bad when two jackals are hunting and the fawn is only a few hours of age and not yet quite firm on its legs.

Male baboons can also become dangerous to gazelle fawns. In searching through an area for edible things, they may encounter a lying-out fawn by chance. On the other hand, a male baboon may start an intentional hunt for a gazelle fawn when it is on its feet. The maternal defense against this enemy is similar to that against jackals. However, in the—admittedly only few—cases I observed, the mothers did not defend their fawns as energetically and persistently, and the young were killed in all these cases. I had the (of course, merely subjective) impression that the gazelle females did not fully "realize" the situation with the baboons.

Very remarkable is the behavior of tommy mothers toward hyenas chasing their fawns. A spotted hyena is much too big to be attacked by a tommy female. However, the mother—and occasionally a second female as well—will dash through between her fleeing fawn and the pursuing hyena. When the pursuit goes over some distance, she does so several times on a zigzag course. It looks as if she were trying to screen the fawn with her body and/or to divert the hyena. Neither I nor any other observer—including my colleague Hans Kruuk, who studied the behavior of hyenas for several years—ever witnessed that these maneuvers of a tommy mother were successful. In most of the cases the fawns were killed, and in the few cases in which a fawn escaped, it appeared quite likely that this would have happened anyway and without the mother's involvement; for example, when the fawn was somewhat older and the hyena had started the hunt too soon from too far a distance.

This behavior of tommy mothers toward hyenas obviously does not fit certain hypotheses and theories on animal behavior, particularly since it does not lead to success in most, if not all, cases. First, it does not neatly agree with the principle of energy conservation, of an "economical" use of an animal's energies. (By the way, there are more facts which hardly can be said to be congruent with this "principle." As only one example, the reader may be reminded of the numerous and certainly very energy-consuming but finally always futile attempts of territorial tommy bucks to prevent females, including nonestrous females, from leaving their territories.) In addition to the fruitlessness of her efforts, a tommy female endangers her own life in the described "diverting" maneuvers, since spotted hyenas do prey upon adult gazelles. Thus, certain theorists cannot understand how phylogenetic evolution could bring about such an unsuitable behavior. And since the theory cannot be in error, it must be Mother Nature who has made the mistake. In short, it is a maladaptation—to give a name to the child. However, when those prescribed and solely beatific hypotheses are replaced by some common sense, an interpretation of the said behavior does not appear to be so far out of any possible reach.

When something has a special meaning in an animal's world (its "Umwelt" in the sense in which the great biologist Jakob von Uexküll used this term), and when this meaning is not a negative (as, e.g., in the case of an enemy) but a positive one (as with respect to food, a territory, a sexual partner, a companion, etc.) resulting in an individual bond, which all is certainly true for a gazelle female with respect to her fawn, this thing may easily become an "object of value" to the animal which is defended when it is in danger of being taken away (in any form)—provided the animal can recognize and realize the situation, which again is obviously true when a gazelle mother sees her fawn being pursued by a hyena. On the other hand, any tommy living in Serengeti has sufficient opportunities to learn that a hyena will prey on adult gazelles as well, although it is not quite as feared by them as some others such as cheetahs, wild dogs, etc. Thus, in the situation under discussion, the tommy mother is under strong pressure to do something for her endangered fawn, but simultaneously she is in (though comparatively moderate) fear of the hyena, which is so hopelessly superior to her in strength that she has no chance for a truly effective defense. In such a desperate situation with such strong but conflicting desires, in men and animals, the most original and extremely strong tendency is at least to stay as close as possible to the endangered "object of value," although this usually does not help anything, and is not very wise, well-adapted, economical, useful, etc. This is rather precisely what a tommy mother does with respect to her fawn pursued by a hyena.

According to another hypothesis on predator-prey relationships, predators largely influence the social behavior of prey animals. Especially, their social organization is said to be "shaped" by the avoidance of enemies, and the gregariousness of some prey species is explained by the fact that "many eyes see more than two"—a truism which quite frequently does not hold up in a specific situation. Also, it must be added immediately that many animals are more conspicuous than one animal alone. Thus, the possible (but by no means sure) advantage that an enemy is spotted sooner by a herd than by a single animal, is outweighed by the (pretty sure) disadvantage that a whole herd may more easily and at a greater distance be detected by a predator than a single animal. It almost goes without saying that adherents of the "many-eyes" hypothesis do not consider this disadvantage. Why and how the social organization of a gregarious prey species—its grouping in all-male herds, female herds, mixed herds, etc.—is supposedly "shaped" by the influence of predators, must be explained by those who hold such opinions. I do not know, and I have never understood it.

All of these speculations stand and fall with the basic assumption of the dominating role of predators in the life of prey animals. Since the African plains animals are frequently quoted as examples of how such prey species have to be on the alert twenty-four hours a day and live to see the next morning only thanks to their permanent readiness to flee, I wanted to learn whether this is true. Doubtless the gazelles in Serengeti offer an excellent opportunity for such a study because a number of predatory species prey upon them, and some of these predators are quite numerous in the national park.

One of Hediger's great merits was to find and emphasize the importance of "flight distance" in animals. Generally, an animal does not necessarily flee as soon as it has seen a predator (or another prospectively dangerous object) but only when the latter has approached to within a certain distance, the flight distance, which may be different in the different species. This could be beautifully observed in the open plains of the Serengeti National Park. Furthermore, I found here that within a given species, the flight distance is not of constant size, but it varies according to several external and internal factors. One concerns the previous experience of the prey animals with the different predatory species. Among all the predators in Serengeti, cheetahs and wild dogs are the most specialized in hunting gazelles and do it most frequently. Consequently, the flight distances of gazelles from these two "arrant enemies" are larger, on average, than from others, such as lions, who also kill gazelles but are at least as frequently, if not more often, after other (bigger) prey animals. The number of predators is also important. Gazelles usually flee from a pack of hyenas at greater distances than from a single hyena. Furthermore, it makes a difference whether the gazelles have recently been heavily disturbed, for instance, by the hunt of a pack of wild dogs. When this has happened, their flight distances from other enemies (including cars, humans, etc.) as well are enlarged for hours.

When a predator is well visible to the gazelles, their flight distance often is surprisingly small—occasionally even with respect to a cheetah, one of their "arrant enemies."

Two further important factors are the speed and the direction taken by a moving enemy. A running predator induces flight at greater distances than one that is walking. And—as always, under otherwise comparable conditions—the gazelles flee earlier when a predator takes a direct course toward them than when (at first) it is moving more or less parallel to them.

I recorded flight distances in gazelles of between one and two miles from a pack of wild dogs that had hunted in a limited area several times, but most of their flight distances are much smaller, sometimes surprisingly small. Even with cheetahs, flight distances are seldom more than six hundred to seven hundred yards, and most are between one hundred and three hundred yards. From lions, gazelles flee at distances of hardly more than four hundred yards, and from there they range down to fifty yards. With hyenas, flight distances of more than three hundred yards are unusual; many of them are between fifty and one hundred yards, but even smaller ones, down to five yards, are not uncommon. Five to thirty yards is also the avoidance distance (they do not run at a gallop but only step aside walking) of adult gazelles toward jackals, and exceptionally they will even let a lion or a cheetah get this close. From wild dogs, however, the smallest flight distance I ever saw was still about one hundred yards.

To avoid misunderstandings: Even in the case of the smallest flight distances mentioned above, it was not that the gazelles had not seen the predators earlier. I am talking here exclusively about cases in which they clearly were aware of the predator's presence for some time before they took flight. It may also be mentioned here that gazelles, like many other game species, when they have spotted a predator at a far distance, may approach it to within about one hundred yards, or even somewhat closer. They watch the enemy intently, uttering alarm calls. When the predator moves, they follow it, keeping the said distance, and thwart its hunt in this way. At least a surprise attack is made impossible. In gazelles, this behavior is shown by single territorial bucks as well as by whole herds bunching together in this situation, and by males as well as by females.

Some people say that prey animals can recognize whether a predator has had enough to eat or whether it is hungry. At least when taken literally, this is very unlikely. What they can see is whether the predator maintains a more or less constant distance or whether it is approaching the prey animals, whether it moves at normal speed or very fast or, vice versa, suspiciously slowly, whether it walks along openly or tries to make use of cover, and so on. In short, they may be able to recognize whether it has hunting intentions. Correspondingly, they may allow it to come relatively close to them, or they may flee at a larger distance.

It is a strange sight to see a lion take its path through a concentration of Thomson's gazelle. As long as it walks calmly and openly, the gazelles do not flee, but give way in front and beside it by stepping away at a quiet pace. Behind it they come together again. Thus, the lion is permanently surrounded by an empty space with a radius of fifty to one hundred fifty yards which is moving with it.

As I said, gazelles overall exhibit larger flight distances from certain predators than from others. This is true for males and females alike. However, with regard to a particular species of predator, adult males exhibit a smaller flight distance on average than the adult females. This spatial difference can also be translated into a temporal difference. When the gazelles become aware of an enemy only after it is already within the flight distance, the males frequently run later than the females. This is most striking in territorial bucks.

When there is a tommy concentration of hundreds or thousands of

KEEPING THE ENEMY UNDER VISUAL CONTROL

Above: A lioness rests in the shade of a tree near a tommy territory. The territorial buck ("Short-tail") has become aware of her presence, approaches her (within the boundaries of his territory), and . . . *Below:* . . . finally beds down face to face with the lioness, keeping her under constant close watch as long as she remains in the vicinity of his territory.

Lioness walking through a tommy herd.

animals in the completely open plains, there are always some territorial bucks among them. However, it may take some time to recognize them in the crowd of females, half-grown young, and nonterritorial males. This changes immediately when a feared predator, such as a cheetah, shows up at the edge of the concentration. After sounding the alarm and watching out, all the gazelles flee—except for the territorial bucks. Highly erect, they stand in their territories and watch the cheetah. Only when the enemy comes closer or even starts running toward them do they flee, too. When I observed such an event, I was always reminded of a deciduous tree in autumn losing its foliage in a heavy gust of wind. Just as the tree's branches become visible all at once, so the "mosaic field" of the territorial bucks shows up within a tommy concentration upon the arrival of a feared predator.

It seemed to me that the differences in flight distances of Thomson's gazelle according to sex, social status, etc., deserved a more systematic and quantitative investigation. Since the animals exhibited the same kind of differences in their flight distances from a car, I drove my Land Rover in the lowest cross-country gear (one-quarter of the speed of normal first gear) directly toward an all-male group or a female herd. I stopped as soon as I saw the first animal run away, walked to the spot where it had been standing at that moment, and measured the distance back to my car with a steel tape measure. In this way, I got only one flight distance for one member of the herd—as it was with single animals, such as territorial bucks, anyway. Altogether, I collected one hundred flight distances from (of course, different) territorial, adult bucks, one hundred from adult bucks in all-male herds, one hundred from adult females in female herds, and one hundred from adolescent males (with horns shorter than their ears), fifty of them in female herds, fifty in all-male herds. The task was most difficult with the solitary wandering adult bucks. So as not to mistake one of them for a territorial buck, which also spend lengthy amounts of time alone, I had to observe each of them for at least one hour. Of course, it was helpful that I knew the location of many territories in the Togoro plains so exactly. On the other hand, solitary wandering males are not encountered very frequently. In short, it took me some time, but eventually I got my one hundred flight distances from them, too.

As expected, the smallest flight distances were found in the territorial tommy bucks. They ranged from not quite five meters (Old Roman!) up to fifty meters, with most between ten and thirty meters. The flight distances of the adult bucks in all-male herds were also pretty small, but they extended up to sixty meters, with more than fifty percent of them between twenty and forty meters. Next, but statistically significantly different, came the adult females in female herds. Their flight distances were between ten and seventy meters, with a well-pronounced peak between forty and fifty meters. The flight distances of the adolescent males in the herds were not significantly different from those of the adult females. The largest flight distances were found in the solitary wandering, nonterritorial, adult males. They ranged from twenty to one hundred forty meters and did not show a clear peak, in that most of them were distributed over the relatively large span from fifty through one hundred meters. Probably this result reflects the fact that these males do not belong to a special social class and are not representative of a homogeneous category. Some of them may be territorial males who have left their territories temporarily, some may be bachelors who have left the all-male groups or the mixed herds in search of a territory but are not territorial as yet, some may be bachelors who entered a territory as members of an all-male group but then were chased away by the owner of the territory in different directions and lost contact with the group in this way, and some of them may be animals separated from a herd by the activities of predators or other external forces.

Comparatively large flight distances, as are typical of the solitary wandering adult bucks, may be interpreted as a sign of considerable insecurity. Very small flight distances indicate the opposite. Thus, the small flight distances of adult tommy bucks in herds and the even smaller ones of territorial bucks may be taken as an indication that staying in his own territory as well as being together with companions in a herd provides a feeling of security to such an animal, whereby, roughly speaking, territory and herd substitute for each other. When an adult male is neither in his territory nor in the company of conspecifics, he is considerably less secure. This was not a new finding, but it was demonstrated particularly well by the comparison of flight distances in adult tommy bucks.

To me, the most interesting point of these quantitative investigations was the confirmation of my previous observations on differences in flight distances according to sex, age, and social status. As described above, these differences do not occur only in flight from a car, but also in flight from predators, although with different dimensions (depending on the species of the predator, etc.). The predators, however, are as dangerous to males as to females, to adolescent males as to adult males, and to territorial bucks as to nonterritorial males in herds. Thus, the differences by sex, age, and social class can hardly be explained as being adaptations to predation. On the contrary, they are incomprehensible in this context. They begin to make sense when we look at intraspecific situations. When a tommy female or an adolescent male is threatened in any form* by an adult buck, the female and the youngster are "well-advised" to react readily and soon, for instance, by withdrawal or flight, because they are hopelessly inferior to an adult male from the outset and cannot take the risk of a fight against him. However, an adult male threatened by a companion is more or less equivalent in strength to the challenger. He can much more easily take the risk of standing and fighting back and does not necessarily have to obey another adult buck's threat as promptly as a female or a young male. This is even more valid for the highly dominant territorial bucks, who, in addition, do not readily leave their territories. Thus, in the intraspecific realm, differences in flight and avoidance distances according to sex, age, and social status are quite understandable. So when the same differences now occur with respect to predators, I see no other interpretation than that an originally intraspecific behavior has found its way into an interspecific situation. In other words, Thomson's gazelle react to predators as if the latter were superdominant and very feared conspecifics. This is

---

*On the behavior of adult tommy bucks toward females, see chapter 4.

rather the opposite of the opinion that the social behavior of these prey animals was "shaped" by the predators.

In general, it simply is not true that the game animals in the African plains live in constant alertness and fear of predators. Many times I have spent an entire day with the herds of Grant's and Thomson's gazelles without seeing one alarm reaction, let alone a flight. When a predator shows up in the surroundings, of course, the gazelles give alarm, and they flee, if necessary. This is not too surprising, after all. However, there is good reason to be surprised about their indifference toward the presence and activities of predators. For instance, when a lion or a cheetah has hunted and killed a tommy, the others flee for two or three hundred yards, stop, and look back. Sometimes they may run another one or two hundred yards, and during the next quarter to half an hour they are more restless than usual. After that, however, they start grazing again; two bucks may threaten and/or fight each other, a territorial buck may start to herd females, and so on, and all this in full sight of the predator eating or carrying away the victim. I could not avoid the impression that, except for the lengthy hunting chases of wild dogs, the activities of predators do not mean much more to the gazelles than a heavy thunderstorm, and that they accept them in much the same way hardened pedestrians accept the car traffic in a big city. Of course, they may pay some attention and stick to certain precautionary measures; however, basically they cannot do much about it and just have to live with it. "Lightning strikes"—one is dead, the others escaped, peace returns, and life goes on, "business as usual." It certainly is only a moderate exaggeration to say that these animals live as if they were alone in the world.

Another quite modern and widely spread view with which I cannot agree concerns stotting, a type of locomotion which occurs in several species but is particularly typical of gazelles. Contrary to most other leaps, stotting gaits are not used to clear (vertical) obstacles. This is, so to speak, a weak point of gazelles and some other open plains animals: they do not commonly jump obstacles, such as fences, but rather go around them or creep through underneath, if possible. In stotting, the animal's hooves usually are only about a foot and a half above the ground. Essentially higher performances, up to five feet, are rare in Grant's and Thomson's gazelles. Apparently, the animal jumps up mainly from the pastern joints and more or less with all four legs simultaneously. During the suspension phase, the legs are loosely stretched downward, the forelegs being beside each other, and the hindlegs

Stotting of a subadult tommy.

beside each other. In landing, the hooves of all four legs touch the ground at almost the same moment, sometimes the hindlegs shortly before the forelegs. Stotting may be repeated several times, resulting in a chain of such actions. A distance of three to five yards can be covered in each stotting jump; however, sometimes the animal may "dance" almost on the spot. Stotting can be contagious (allelomimetic effect) to conspecifics, not least in the running plays of juvenile gazelles. Generally, it is more frequently seen in young animals than in adults, and in flight from certain predators, such as wild dogs and hyenas, more regularly than in flight from others, such as lions and cheetahs. If stotting occurs in a flight, it is predominantly at the beginning and/or the end. Thus, a gazelle may initiate the flight by stotting jumps but then continue it at a flat gallop. When later the enemy has given up and ceased the pursuit, the gazelle may end the flight with several stotting movements.

In the 1970s, a special concept of "altruistic behavior" was developed in sociobiology. Of course, there is altruistic behavior in animals—for instance, when a tommy mother defends her fawn against jackals. However, provided that I understood them correctly, sociobiologists inflated this term beyond its more or less motivational meaning in common usage (an individual's action is aimed to support a partner) to a merely functional meaning (one individual's action turns out to be advantageous to another regardless of whether this was intended or whether the action happened to have this effect more or less by chance). Furthermore, "altruism" in this broadened sense was linked with certain ideas on the preservation of an individual's genes in its descendants and close relatives. In short, what sociobiologists apparently had in mind when talking about "altruistic behavior" was any behavior which is not advantageous to the performing individual (it may even be disadvantageous or fatal to it), but which appears to be genetically beneficial to its offspring. They included the stotting of gazelles as an example in the concept; obviously they believed that a gazelle warns and alarms offspring and relatives but also attracts predators' attention and pursuit with its striking stotting actions. Thus, the performer jeopardizes its own life by stotting; however, this behavior is of genetic benefit to its offspring. Consequently, for instance, E. O. Wilson in his representative book *Sociobiology* declared stotting to be an altruistic behavior based on kinship.

Certainly these hypotheses promise a deep insight into the ways of Mother Nature. But what is true about them in the light of facts?

At first, we may ask what evidence is available for the assumption that stotting may jeopardize the life of a gazelle fleeing from a predator. I know of only one "argument." In a few cases in which a gazelle initiated a flight by stotting before moving into a flat gallop, and was caught by the predator, certain observers have expressed the opinion that the animal might have had a better chance to escape had it started the flight with the gallop from the beginning. However, this is an absolutely subjective impression on the part of the human observer, and, since the animal is dead, there is no way to prove or disprove whether it truly would have been able to escape had it taken off at a gallop immediately. Most certainly, this cannot be considered a fact, and the best that can be said about the whole question is that it remains completely undecided.

The next question concerns kinship. With respect to the hypotheses under discussion, the relatively most favorable situation exists in the female herds and the mixed herds. Male fawns stay with their mothers until they are at least five months old, and female fawns considerably longer. Thus, it is not unusual for close relatives to be together in such a herd. However, most female herds of Thomson's gazelle number between twenty and fifty, and herds of more than one hundred do occur; mixed herds may even be much larger. Thus, the number of animals related to each other by kinship constitutes only a small proportion of a herd. Even leaving aside all the objections which must be raised to the signal character of stotting and its alarming effect (see below), and presuming that a female would indeed warn the others and attract the predators' pursuit by stotting, her behavior and possible sacrifice would benefit by far more nonrelatives than relatives. Thus, when the presumed "altruism" is focused on kinship, it appears to be a rather dubious matter.

Of course, stotting may also occur in all-male groups. The possibility that, for instance, father and son, or older and younger (half-) brothers, etc., may be united in such a herd cannot be excluded. Taking into account, however, the gestation period, as well as the facts that a male fawn usually does not join an all-male group before the age of five to six months, that there have been moves and migrations during all this time, and that the herds of gazelles are unstable units which easily split and amalgamate, the probability that individuals related to each other by kinship are united in the same all-male group is pretty remote.

The last and most decisive question concerns the evidence we have to consider stotting as an alarm signal. I know of only one situation in which this function may be attributed to it with good probability. When a lying-out gazelle fawn is disturbed by something to the point that it jumps up and runs away, it often initiates the flight by stotting. Although a gazelle mother usually keeps her fawn's location under visual control and thus becomes aware anyway when something out of the ordinary occurs, it is not unlikely that the fawn's stotting may have a reinforcing effect and make it perfectly clear to the mother that her young now is in danger. However, this situation is far different from the things required by the hypotheses of sociobiologists! When, for instance, a lying-out fawn is detected by a jackal and is attempting to flee, it does not need to attract the predator's attention and pursuit any more, and its stotting is not altruistic at all. On the contrary, it calls for the mother's help.

When a gazelle in a herd initiates its flight from a predator by stotting, the alarm function of this behavior is no more than a matter of "maybe or maybe not." Having become aware of an enemy at a distance, usually a gazelle will first watch it closely and utter the alarm call by nose—in Grant's gazelle, a deep and relatively loud "kwoof"; in Thomson's gazelle, a rather soft "kwiff." Thus, the others are already alarmed before the animal begins stotting. Moreover, when a herd is attacked by a predator, many or even all of the gazelles may begin stotting. One could speculate that this might have a confusing effect upon the enemy in such cases.* But who warns whom, and who does something to whose benefit, when all the animals do the same simultaneously?

While an alarming effect of stotting could perhaps still be taken into consideration as a (remote) possibility in the situation above, there are other situations in which stotting clearly does not have any such effect, and certainly does not have anything to do with altruism.

When gazelles end a flight by stotting, the danger is over. Either the predator has ceased hunting, or it has killed a gazelle. An alarm comes too late in this case, the stotting animal does not attract the predator's attention and pursuit, and there is nothing altruistic about it.

Also, the frequent stotting of young gazelles during running plays exhibits neither an alarm function nor any sign of altruism. At best, the adults, including the fawns' mothers, may keep an eye on the stotting young, but often they continue grazing as if nothing had happened.

Finally, in some (admittedly relatively rare) cases, stotting is even linked with pursuit—the opposite of flight. For instance, when two tommy bucks are fighting, and the weaker one suddenly turns to flee (without stotting), the victor, after being startled briefly, sometimes will initiate the pursuit by stotting.

I may mention here that I pointed out the above in 1969, and this paper was known to the said theorists, since they quoted it. However, they carefully and completely ignored everything which did not fit their hypotheses. In short, when coming back to the initial question of what is correct in the sociobiologists' statements on stotting, in the light of the facts concerning kinship, altruism, and alarm effect of this behavior in Thomson's gazelle, the answer can be given in one simple word: nothing.

There are a couple of additional theories which are closer to the truth than the ones discussed above but are still debatable. The first states that prey species benefit from the selection exercised by predators. This view is usually interpreted to mean that healthy, strong, and fit animals do not easily fall victim to predation, and that it is predominantly weak, sick, or somehow maladapted individuals that are taken by predators. It is certainly correct to say that any animal that does

*See chapter 1.

not look normal or is behaving unusually may easily attract the attention of beasts of prey, and if it cannot flee at normal speed, it will be caught and killed particularly easily. It must be added, however, that this "category" includes animals which are only temporarily in less than optimal condition and will recover soon if left in peace. For instance, when a gazelle has a slight leg sprain, which with some bad luck can happen to the fittest animal, its limping deviates from normal locomotion and thus is striking. Moreover, this animal cannot run as fast as the others. If it is spotted in this state by a hunting hyena, the rest is self-explanatory. Also, highly pregnant females may sometimes move a little differently and not be able to run as fast as the others, and, above all, giving birth is a process that deviates from the everyday occurrence and can definitely attract the attention of roaming predators. Although a tommy female can still gallop with the neonate hanging halfway out of her vagina, she certainly cannot run as fast as at other times. Thus, she may be said to be "lucky" if the fawn drops down during her flight and the pursuer kills and eats the neonate but lets the female run, as I witnessed twice. Events such as these have little, if anything, to do with selection and with deletion of unfit individuals.

It is absolutely improbable that in an area such as Serengeti, the numbers of sick animals are sufficient to feed the numerous predators—except when an epidemic disease breaks out, which fortunately does not happen very often. Consequently, the predators simply must kill healthy prey animals in most cases. Here, theorists sometimes raise the objection that an animal could be weak or poorly adapted in bodily or behavioral regards without being sick or strikingly different from others, so that a human observer cannot recognize its weaknesses, and that it is such animals which are predominantly killed by the predators. I can fully confirm that a human observer cannot recognize the postulated weaknesses in the animals. For this very reason, however, I am also convinced that this hypothesis is mere speculation which is beyond proof or disproof, and therefore without any value. Which individual out of a herd is killed, whether a fleeing antelope manages to escape from a pursuing predator or is caught and killed, this and more depends so much on circumstances and casual conditions that one can speak only of the good luck or the bad luck of such an animal.

For example, I once observed a herd of Grant's gazelle moving through a high-grass area. (In contrast to the much smaller tommies, Grant's gazelle commonly move through high-grass areas as peacefully as through the short-grass plains.) At the head of the group, three adult females walked closely one behind the other. After an interval of about twenty yards, the rest of the herd followed in file formation. Suddenly, a lioness arose from the high grass right in front of the three lead females. These leaped high up and sideways as if launched by catapults. In the same moment, the lioness jumped forward and, with the anterior part of her body erect, beat the air with her forepaws like the sails of a windmill. She angled down on one of the Grant's females from the leap. The two others escaped. Who can seriously maintain that these two were more viable and fitter than the victim—or the other way around?

To cite only one more example, the reader may recall the way that jackals zigzag through an area in search of lying-out gazelle fawns (a hyena may do the same). When a jackal misses a fawn at a distance of only four to five yards, he usually does not become aware of it. However, when a fawn is directly in the jackal's path, of course he will detect it. As described, a hunt will ensue, and the mother will come to her baby's defense, whereby the jackals have at least a fifty percent chance to catch the fawn. When bedding down for lying out, even the fittest fawn cannot foresee which course a jackal will take hours later. Thus, whether such a fawn rests right on the jackal's course or a few yards off of it is a matter of mere chance, of good or bad luck. With my very best will, I cannot see any "selection" or "survival of the fittest" here.

On the whole, my years in the field taught me that in the life of animals in general and with respect to predator-prey relationships in particular, casual events play a much greater role than the human mind, eagerly searching for reasons, purposes, usefulness, principles, and insights, often likes to accept.

Another widespread opinion is that predators kill their prey quickly so that the victim's death is easy. Of course, there are cases in which this is true. A somewhat smaller prey, such as a gazelle, may be killed instantly when a big predator, such as a lion, knocks it down by paw. According to the "great old man" in cat behavior, Paul Leyhausen, all the felid (cat-related) species can kill their prey by biting its nape. As far as the prey's nape is not too massive, the predator's canines glide more or less automatically into the interval between two cervical vertebrae and hit the spinal cord so that the victim is killed immediately. Also, when a hyena grasps a gazelle fawn obliquely over its back, the hyena's powerful set of teeth crush the fawn's spine in one stroke. In all such cases, the death of the prey animal definitely is instantaneous. On the other hand, especially cheetahs and leopards will sometimes "play" with a gazelle fawn like a cat with a mouse, in which case the fawn's death is not fast and easy. This is even more true when a male baboon "takes apart" a fawn while it is still alive.

In addition to the nape-bite, the big felids have at least two other methods of killing prey. As George Schaller found out, lions in particular may bite the snout of bigger prey, such as wildebeest, and suffocate the animal. As compared to the nape-bite, this means a slower and more agonizing death for the victim. The other method is to bite the prey animal's throat and strangulate it, which is practiced primarily by cheetahs on gazelles. This throat-bite immobilizes the prey immediately. When observing such scenes, I initially presumed it was dead. Often, however, this is not true. The animal may be only unconscious, or simply immobilized by the shock of being caught, and it may be quite a while before it dies.

My younger colleague Morris Gosling, who studied the behavior of kongoni in the Nairobi National Park, had an impressive experience in this regard. Once he visited me in the Serengeti National Park for a couple of days. In talking with him, I mentioned that I had raised several gazelle fawns and a blesbok calf in zoological gardens in previous years. A few weeks later, I received a letter from him which began with the words: "Help! I am a mother." One day in the Nairobi National Park, Gosling observed a cheetah chase and catch a kongoni calf and strangulate it by biting its throat. Morris had long wanted to take a closer look at certain anatomical peculiarities of kongoni in a freshly killed animal, but he did not wish to shoot an animal for this purpose. Now he thought a good opportunity had come to satisfy his curiosity. He scared away the cheetah, took the dead calf, and threw it in the back of his Land Rover. Then he drove back to his quarters at the Nairobi university. Right in the streets of the city of Nairobi, however, the "carcass" behind him came back to life!

In a sense, the prey animals killed by lions, leopards, and cheetahs are "fortunate" victims. The end is much worse for those killed by wild dogs, hyenas, and—predominantly for young animals—by the two jackal species. Besides the fact that wild dogs are superior to the prey animals not so much in speed but in persistence, so that there usually is a comparatively long chase during which the prey animal has time to experience all states of fear, these predators do not utilize a special bite to kill their prey. They bite those parts of the victim's body which are next to them, and since they come from behind when pursuing a fleeing animal, these commonly are its thighs or its inguinal region, where a bite is not immediately fatal. If a pack of wild dogs is involved, death still occurs relatively quickly in a somewhat smaller prey such as a tommy. After one dog has caught the gazelle, the others may bite into it from all sides so that it rather literally is torn to pieces. However, with a bigger prey—but also with a small one when only one or two such predators participate in the hunt—wild dogs, hyenas, and, with fawns and calves, jackals will eat from the living animal, and it may take as long as twenty minutes for it to die. While wild dogs or hyenas are biting into their bodies and pulling out their intestines, wildebeest, zebra, kongoni, and others often remain standing, immobilized in full shock, until eventually they break down. They "watch their own execution," as Richard D. Estes phrased it plastically.

The reader may perhaps understand now why I also have certain reservations with respect to the talk of an instant and easy death for the prey animals killed by predators. Although in the course of four years at African national parks I witnessed more than sixty successful hunts by different beasts of prey, these always were casual observations. I did not make special efforts to see such scenes. For one, to see more of them, I could not have remained with the herds of gazelles and other antelopes, but it would have been necessary to stay with the predators and accompany them on their moves. This was not my research task. Moreover, I do not enjoy watching an animal's misery and dying. I think this requires a somewhat sadistic vein, which seems to be present in some people but is lacking in me—and I see no reason to apologize for it. I admit that in all the cases of predation I witnessed, I breathed easier again when everything was over and the victim was dead. Only when watching a cheetah hunting for gazelles did the marvelous speed of this elegant predator always fascinate me.

When gazelles become aware of a cheetah at a distance of more than two hundred yards, the predator hardly has a chance. In the shortgrass plains, a cheetah seldom can approach them closer than one hundred yards without being detected, but when it has made it to about this distance, its chances are quite good. Now the gazelles have seen it; they give their alarm calls and crane their necks in its direction. The cheetah no longer attempts to stalk but stands or sits right out in the open. Hunter and hunted face each other. Very abruptly, the cheetah will start at a gallop, and at the same moment (or is it a second before or after it? I do not dare decide this question), the gazelles will turn and flee. They run with all their might but, in comparison to the cheetah, it looks as if they are standing still. Often it is after one definite animal from the beginning, possibly the one which first turned to flee and thus released its pursuit (again, I cannot say for sure whether this is so). The cheetah may pass other gazelles, or one of them may cross right in front of it, or, for a moment, a gazelle may even run side by side with it, but it pays no attention to them. When the cheetah is very close to its chosen victim,

Cheetah hunting for Thomson's gazelle.

the gazelle may double like a hare several times. Although the cheetah may lose track for a moment, I never saw that this desperate maneuver led to a final success. The end of the hunt usually disappears from the observer's sight in a cloud of dust. Only on four occasions could I better watch what was going on. Upon catching up with the victim, the cheetah throws the anterior part of its body upward and beats its paw on the gazelle's back or between its hindlegs—which of these two alternatives is true, I never could recognize for sure. In the next moment, the gazelle rolls over on the ground. The cheetah flings itself on it and bites its throat.

There is much sparkling life and much bitter dying on the African plains. But death does not govern life in the wild, though it is always present somewhere in the background and may show up any time. I deeply felt that it is precisely the occurrence of life in the presence of death that distinguishes the nature of freedom in the wild, and, before falling asleep in my sleeping bag in my car, I sometimes tried to imagine which horror might capture our politicians and social philosophers who babble so drolly about "freedom" if they were exposed to this freedom—the only one which deserves the name.

8 Mzee—The Old Man

In 1974/75, I was again in the Serengeti National Park for one year. In addition to continuing my earlier studies on Grant's and Thomson's gazelles, I took a closer look at several further antelope species which I previously had observed rather occasionally. The oryx antelope was one of these.

Oryx antelope are—or at least were—widely distributed over Africa and Arabia in several geographic forms. Some scientists consider them to be species, others speak of subspecies. The fringe-eared oryx inhabits southern Kenya and northern Tanzania, including the Serengeti area. It represents the southern variant of the East African oryx or beisa antelope, which ranges up to Abyssinia and Eritrea in the north. The primary differences between the fringe-eared oryx and the more northern animals are a somewhat darker, more brownish color of the coat and the long, black hair tufts at the tips of the ears.

When Michael and Bernhard Grzimek conducted the first systematic game count in the Serengeti area in 1956, they reported one hundred fifteen oryx antelopes, mainly in the Salai plains between Olduvai Gorge and Lake Natron. However, in 1959 the boundaries of the Serengeti National Park were changed, and this area was cut out of it. After this change, oryx antelope were no longer considered to be among the game species of the park. Therefore, it caused some excitement among the "old-timers" of Serengeti when I and a companion spotted a single oryx bull near the Princess Kopjes, east of Seronera, in 1965. As far as I know, this was the first sighting of an oryx within the new park boundaries. The next year, a single oryx was again seen at about the same place by other observers. No sightings are known to me from the following three years. From 1969 through 1972, pilots of the Serengeti Research Institute conducted monthly reconnaissance flights over the whole of the park and the adjacent areas, and they saw oryx antelopes several times. I went through their records and mapped their oryx sightings. As expected, many of them were outside the national park, in the area between Olduvai Gorge and Lake Natron. To my surprise, however, there were also several sightings at Banagi and in the "Corridor," where occurrences of oryx antelopes were rather unlikely. Eventually it dawned on me that these were the home ranges of the very few roan herds in Serengeti. Apparently, the pilots had mistaken roan for oryx antelope. (I may add here, however, that later Richard D. Estes informed me that he once had seen an oryx halfway between Seronera and Banagi—and Estes certainly had not confused oryx and roan antelope.)

Apart from those at least somewhat dubious cases, there were only three sightings of oryx within the park's boundaries from 1969 through 1972. Particularly famous became a single oryx cow who was present in the area around Naabi Hill in the south of the park for several days. Game wardens and research scientists made a pilgrimage there to see *the* oryx of Serengeti. Beginning in 1973, the number of sightings within the park increased. At first, Dr. Tumaini Mcharo—a Mdshagga* who had completed his academic training at a university in the United States, where he had also taken, among others, my lectures, and who now was the director of the Serengeti Research Institute—and a few other pilots recorded oryx antelopes singly or in small groups during their reconnaissance flights in the southeastern regions of the park. Later, scientists working there on the ground also spotted oryx antelopes several times. Obviously, these animals were moving increasingly but irregularly from the adjacent Salai plains into the national park. What caused this "invasion" was and remained unknown.

---

*Wadshagga (singular Mdshagga) are a tribe near Moshi, Tanzania.

Oryx roaming the Serengeti plains.

This was the situation when I arrived again in Serengeti in May 1974. Of course, I wanted very much to observe the oryx antelopes on their interesting moves through the Serengeti plains. However, I also had numerous other objectives on my research program. Since, for the time being, I was working in a section of the national park where the occurrence of oryx was hardly to be expected, I asked the game wardens and my colleagues to inform me if anybody happened to see one. Late in August, Tumaini Mcharo returned from one of his flights; he had seen an oryx herd of fifteen animals in the southeastern plains. This was the largest herd recorded inside the park up to then. (Later, even larger herds were observed.) In his defense, I have to state that he immediately tried to inform me; however, I was not at home—as may easily happen with a good field researcher. So I got the message only after three days. Since it was to be expected that the oryx had meanwhile moved more or less farther away, I decided to drive along the park's eastern boundary from its center toward the south.

In those days, tourists seldom came to this area, and it also was not often frequented by game wardens and the other scientists. It is an area of breathtaking vastness. I owe to it one of the greatest experiences of landscape and space which I ever had in my life. If, however, something goes wrong out there—for instance, if the car breaks down or gets seriously stuck—you are very much on your own, and you can prove that a man still has his value in the field. There are numerous and often quite remarkable holes in the ground, the entrances or exits of subterrestrial dens, usually established initially by aardvarks but later often used and enlarged by other animals, predominantly hyenas. Furthermore, these short-grass plains are intersected by river beds that contain no water during most of the year, which I had to cross or to follow in a search of oryx, who quite frequently stay on the bottom of such a korongo for grazing so that, at best, just the tips of their horns may protrude over its rim. As I mentioned in an earlier chapter, there are groups of kopjes distributed over these plains.

Since the oryx may easily stand behind such a kopje, I had to drive around each of them. Unintentionally, I chased a tommy buck out of his territory so that he ran into the territory of his neighbor, who immediately charged the involuntary intruder, and the two engaged in a marvelous fight. Right in front of my car, a gazelle fawn struggled up from a big hole in the ground. That tommy fawns may lie out in the entrance to an apparent hyena den was news to me, although I had been studying the behavior of this species for almost three years. Learning never ends. . . . Several times I saw honey badgers, once a bush cat, once a serval. Two black-backed jackals were pestering a golden jackal. In short, it was never boring—but I could not see oryx anywhere. With the sinking sun, I drove around a kopje—and there in front of me were four oryx antelopes. I had only enough time to see that they were one bull and three cows and that they tolerated the approach of my car to within about sixty yards. Thus, they were by no means as shy as East African oryx are often said to be. Then it grew dark, and I went to sleep in the back of my car, as usual.

Next morning, I was up with the first light, but I could not see anything of "my" oryx. I climbed the kopje, and with the binoculars at my eyes, I could recognize them as tiny, moving specks in the plains far to the south. Half an hour later, I had caught up with them. They moved slowly, grazing and pacing, in front of me, and they led me straight to another eleven oryx. They united with this group without complications. Obviously, these were the fifteen oryx antelopes that Tumaini had sighted from his airplane.

The herd comprised six bulls and nine cows, all of them adult animals. One of the bulls impressed me as being stronger and more massive in body and neck than the others. In my notes I named him "Mzee," without knowing as yet how much on the mark I was. In Kiswahili, "mzee" means "old man," and in the mouths of East Africans, who have great respect for aged people, this is a very honorable title.

Up until about nine o'clock A.M., the animals remained at this place,

A dominant oryx bull approaches a subordinate, who responds by lowering his head.

mostly grazing; sometimes one of them bedded down for a little rest. Then, however, they formed a file, one after the other, and moved ahead in a southerly direction, Mzee as the last one in the file. During this march, two bulls started a more or less playful fight and remained behind the others. When the herd was more than two hundred yards away from them, the foremost walking animals deviated from the previous course at a right angle. However, Mzee did not seem to approve. He passed the marching file at a gallop, stopped the animals at its head by blocking their path in broadside position, and chased them back. Apparently only then did he become aware that the two bulls struggling with each other had remained behind. He left the herd and galloped the whole way back. Shortly before arriving at the fighters, he slowed down to a walk. The two bulls ceased fighting, turned toward him, and started grazing. Mzee approached closer until he stood in reverse-parallel position beside one of them. He assumed an erect posture and angled his horns sideward toward the younger bull, who stopped grazing and turned his horns away from Mzee, keeping his head low. He more or less

Sideward angling of the horns and "pointing" ear of a dominant oryx bull in reverse-parallel position toward a subordinate who is withdrawing with lowered head and tail-wave.

The dominant oryx bull directs a symbolic horn blow in slow motion toward a subordinate, who turns his horns away in submission.

turned his throat toward Mzee. For a moment he remained in this submissive posture, then he fled at a gallop toward the distant herd. His former opponent ran with him. Mzee followed them, speeding them up several times on the way.

Upon their arrival at the herd, these two bulls first brought several herd members between them and Mzee—as "security factors." Again the group formed a file and moved ahead, with Mzee as the "shepherd" at the rear. And again the foremost animals turned in the direction which Mzee did not like, and now I could see why. About three hundred yards away in this direction, another oryx bull had shown up who was about as strong and massive as Mzee but who obviously did not belong to this herd. Mzee charged him at a gallop. The other turned and fled, however, only for a short distance. Then he stopped, turned, and waited for Mzee, who now approached him at a walk. The two stood, facing semi-obliquely toward each other, in highly erect postures and turned their heads to the side, away from each other. Suddenly they attacked simultaneously with all their force. They beat with the length of their horns forward-downward, twisting their heads somewhat so that their horns and/or foreheads hit each other crosswise. The exchange of heavy blows of this type alternated with phases in which they pressed against each other with horns

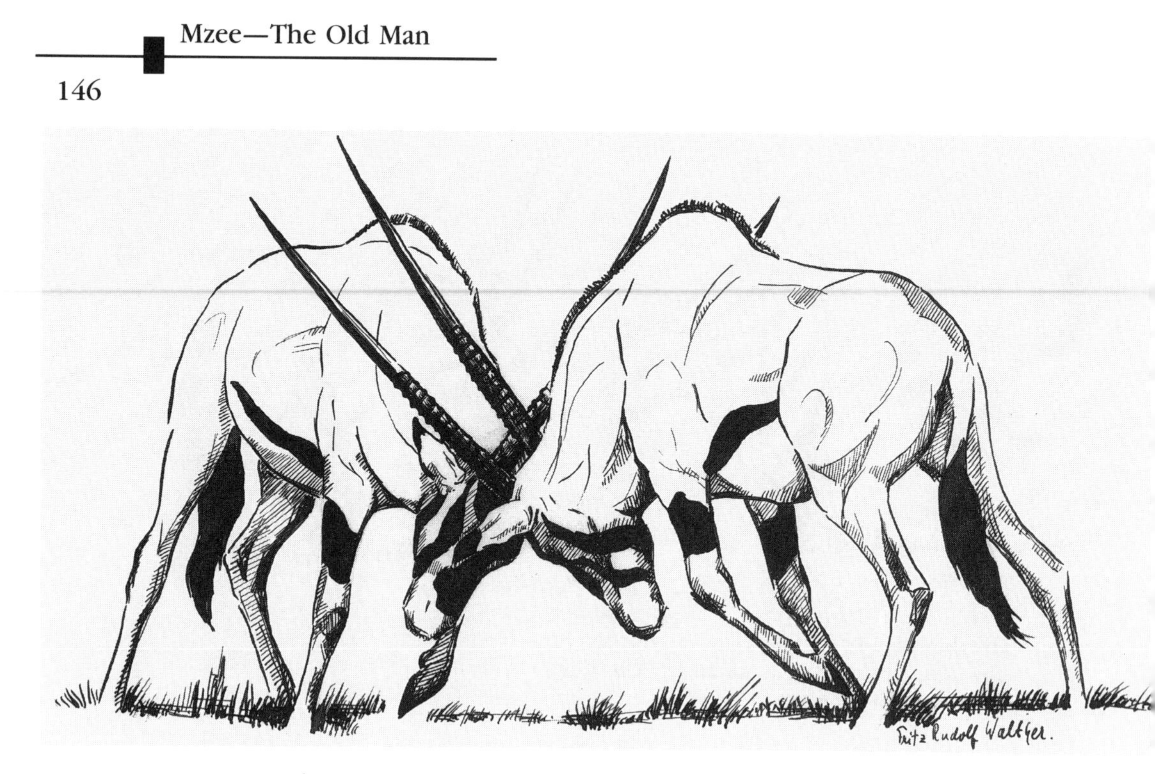

In reciprocal horn blows, the horns and/or the foreheads of the adversaries hit each other crosswise.

interlocked. In a powerful forward thrust, Mzee pushed his adversary back over several yards, but in the next moment he was moved back by the other about the same distance. Both stopped fighting, stepped backward, grazed—and started fighting again. They remained on their legs the entire time and did not drop down to their knees (carpal joints), which otherwise is not uncommon in fighting oryx. Although this fight doubtless was serious, there was no attempt or intention recognizable to gore the opponent's flank. After a pause during which they grazed or walked side by side, the fighters turned to face each other either in erect posture or—now more frequently—in head-low posture. Then they touched each other's noses, "taking measure," before they interlocked their horns again and pressed against each other.

The fight had lasted more than a quarter of an hour already, and the pauses between "rounds" had become longer and longer. Sometimes now one of the opponents dropped to his knees and gored the ground with his horns, or he scraped the ground with a front leg so that the

Fighting in parallel position, the opponents press and push against each other with their shoulders.

Fritz Rudolf Walther.

Fighting bulls frontally press against each other with horns crossed.

clods flew, or he defecated, crouching so deeply that his anal region almost touched the ground. After such a pause with a long parallel march and a short parallel gallop, Mzee simply left his adversary standing there, stepped toward his herd, and drove it away from the rival. The latter stood for a few minutes at the spot where Mzee had left him, then walked off in the opposite direction.

Meanwhile, once again, two of the younger bulls were involved in a lengthy fight, this time right within the herd. In comparison to the fight of the two old bulls, this one appeared almost "cozy." Without obviously being defeated, one of the combatants suddenly turned to flee, immediately pursued by the other. Twice the chase went through the whole herd. Then a third bull—with one horn not as straight as the other but curving a little backward, so I named him "Curved Horn"—intervened. He stepped between the pursued bull and the pursuer, fought the latter pretty vehemently, and eventually chased him off.

He probably would have chased him for quite a while if Mzee had not very literally crossed his path. After his serious opponent, the strange heavy bull, had disappeared, Mzee had bedded down and did not seem to care about what the other bulls were doing. Now, he rose and stepped between the pursued one and the pursuer. He blocked Curved Horn's path in broadside position. Curved Horn stopped from full gallop, stood, and lowered his head to the ground. Mzee erected and turned his head with presented horns toward Curved Horn. Slowly he turned into a frontal position, and slowly, very slowly, he lowered his head toward him—a horn blow in slow motion. In response, Curved Horn leaned his horns back toward his nape. This corresponds to a parrying movement in a fight by which the defender screens his body and catches the attacker's blow with his horns. This time, however, both offensive and defensive movements remained purely "symbolic" performances, in that the opponents stood apart and did not touch each other. Eventually Curved Horn could bear it no longer. Abruptly, he turned sideways and took off at a gallop.

Meanwhile, it was high noon. The herd's activity had changed from moving to grazing, and from grazing to lying. Only Mzee and one of the cows were still on their feet. These two stood in reverse-parallel position, the bull in normal or even slightly erect attitude, the cow with lowered head. Slowly they circled around each other, a behavior resembling the encounters of Mzee with subordinate bulls, but the distance between the partners was smaller, and the cow did not flee. Eventually Mzee raised his foreleg in a long, emphatic, and rather stiff-legged step (a ritualized foreleg kick) in the direction of the cow's hindlegs. She urinated, he sniffed at the streaming urine and performed Flehmen (lip-curl). After that, he left the cow and bedded down.

The herd rested for about an hour—just time enough for me to note down the grouping of the resting animals, the distances between them, and their orientation toward each other, to the sun, and to the wind. Then Mzee got up. With erect head he passed along the resting herd members. Turning his head slightly away, he "pointed" with one ear to one of the bulls. The black hair tuft at the tip of the ear of a "fringe-eared oryx" makes this "pointing" particularly striking. The resting bull rose, and shortly thereafter, Mzee repeated the procedure with a second and a third bull. The other members of the group now also got up, one after the other. Mzee had encounters with several bulls in which each of them responded to his threat and dominance displays with a head-low posture, expressing a mixture of submission and readiness for defense. After a shorter or longer while, each of them walked off. Sometimes Mzee only looked after the withdrawing partner; sometimes he followed him, walking a short distance; and sometimes he chased him at a gallop. Thus, the restlessness within the herd increased. After rising from resting, most of the animals grazed at first. Then they formed a file again and began to march, with Mzee, as usual, bringing up the rear.

The move led in a southerly direction as it had in the morning. The antelopes marched uninterruptedly and fluently, and I had to "get myself in gear." First I drove about three hundred yards sideways; then I passed the herd at this distance, drove ahead for at least half a mile, stopped, and waited for them to come. This is the best method for keeping migratory plains animals under observation without influencing the speed and direction of their moves. The difficulty lies in predicting which course they will take, since, although they usually maintain the same general direction, they may deviate somewhat from a straight course. However, I got some training in this regard by observing migratory gazelles, and thus was able to correctly predict the route of this oryx herd most of the time.

The march proceeded briskly, although not without incident. From time to time a herd member stopped walking and started grazing. When Mzee approached—with his head slightly turned away and pointing his ear—the lingering animal moved ahead again. Twice it happened that herd members escaped Mzee's attention and remained behind. In both cases, Mzee became aware of their absence after a march of over two hundred to three hundred yards. Only he stopped, turned, walked back to the stragglers, brought them to the point by threat and dominance displays, and made them follow the herd in a hurry. At the beginning,

the lead position in the marching file was taken by several herd members alternately. Later on, one and the same cow kept this position, immediately followed by the bull Curved Horn, who occasionally drove her and sped her up. "Pulled" by these two "outriders" and "pushed" by Mzee from the rear, the herd moved ahead, one after the other.

Shortly before sundown, Curved Horn suddenly stopped marching and placed himself with his flank toward the following animals, blocking their path. Of course, they could walk around him, but he interrupted the flow of the move, the more so since he threatened each member of the group when it reached him, and he even made some of them withdraw. Finally, Mzee approached, and he, of course, was not "thinking" of a withdrawal. He moved into a reverse-parallel position with Curved Horn, took an erect stand, and angled his horns toward him. Curved Horn, from his previously erect dominance posture, immediately "folded down" to the submissive-defensive head-low posture, turned around, and galloped in the "demanded" direction. Under Mzee's threat and dominance displays, the herd resumed its file formation and marched on.

In the last daylight, at about seven o'clock P.M., I could once more place myself well in front of the herd, and a short time later I had the joy of seeing these marvelous creatures approach and pass my car at a distance of hardly twenty yards. I did not dare to switch on the lights of the vehicle because I did not know how oryx might respond to the light. The herd had moved more than six miles within the last two hours, and there was no indication of an early end to this march. Unfortunately, moonlight would not be available for a few hours. Thus, for better or for worse, all I could do was follow the animals in the dark.

For half an hour, the shapes of the marching oryx antelopes appeared in front of me like shadows from the dark. Once I could even identify Mzee by his straight, long horns. Then it happened: the right front wheel of my car got stuck firmly and deeply in a big hole, the

Above: A bull has remained about two hundred yards behind the moving herd. "Mzee" has walked back to him, has passed him, and is now blocking his path in broadside position (lateral T-position) to prevent him from straying farther away.

Center: The straggler stands in the typical defensive-submissive head-low posture. "Mzee" turns slowly toward him, threatening by (slightly) high presentation of horns.

Below: The straggler, with lowered head, turns and follows the herd. "Mzee" drives him ahead and speeds him up.

Moving oryx herd with straggler.

entrance of a hyena's den. To hell with these beasts! This thought was neither particularly wise nor objective, but it was human. Even with four-wheel drive and cross-country gear, I could not get free. I had to get out of the car, dig, and use the jack. The oryx antelopes had long since disappeared in the dark at full gallop. Had I been "in their shoes," I would have done the same upon this hullabaloo. Eventually my car was ready for action again, but by then it was out of the question to follow and find the oryx herd that night. On the other hand, all the film I had with me was exposed, the tapes were filled, and the day had been unique. Gratefully and contentedly, I leaned back in my seat and lit my pipe—it was only the second one for which I had found time all day.

While I was smoking, some "wise" thoughts crossed my mind. That antelope males herd the females of their harems and more or less successfully make them move in a definite direction, this is not particularly rare and can easily be observed, for instance, in impala. Also, that adult males in all-male groups or in mixed herds can speed up the changes from resting to grazing or moving and can keep a migration going by threatening, fighting, and chasing the others is not unusual and is pretty familiar to me from my observations of gazelles. This comparative background, however, made it very clear to me how different were the things I had seen that day in the oryx. Mzee did not herd only the females but also and even particularly the bulls, since at least some of them tended to separate from the group. Also, for instance, in a migratory herd of Thomson's gazelle, an adult buck may act or react when the animal right in front of him slows down, deviates from the course, etc., but he does not leave the herd and walk back two hundred yards or more to pick up a straggler, as Mzee very clearly did several times. Finally, here only one individual, Mzee, was responsible for everything going on in the herd, and became its "very life and soul" in this way. This I never experienced in any other antelope species I observed. Is this a general behavior in oryx, or does it occur only when a herd is moving through a vast area not particularly familiar to the animals? The deeply crouched defecation posture of dominant adult bulls could have something to do with marking behavior, although not necessarily with marking a territory, since Mzee quite certainly was not territorial. But the single bull that he had fought in the morning possibly could have been territorial or on the way to become it. And what was the intention of the bull Curved Horn when he tried to stop the move of the herd in the evening? How can it generally be explained that expressive movements and postures, such as threat and dominance displays, have the same power and effect upon the recipients as physical actions, such as blows or pushes? Can we in such cases exclude the presence and the effectiveness of psychological factors, of a mind and of mental processes—provided we take things seriously and do not merely try to find substituting vocables for the words "mind" and "mental" disliked by so many scientists in discussions on animal behavior?

Such and similar thoughts and questions came to my mind, and—not for the first time—I came to the conclusion that we do not know too much, either about the animals or about ourselves. Will we ever come to know everything? And if we did have all the facts, would we be able to understand and interpret them correctly? Be that as it may. What remains is the joy of animal life, and the gratitude that such things still exist on earth, and that there is still a science the substance of which is made up of such events of true life.

The stars of the clear African sky twinkled above me. "Zwei Dinge erfüllen das Gemüt mit immer neuer und zunehmender Bewunderung, je öfter und anhaltender sich das Nachdenken damit beschäftigt: der gestirnte Himmel über mir und das moralische Gesetz in mir" ("There are two things which again and again fill the mind with new and increasing admiration, the more frequently and the more continuously the thinker ponders them: the starry sky above me and the moral law inside me"). Well, the nice old philosopher Immanuel Kant never spent much time outside of Königsberg, and as far as I know, he never even tried to see the moose in his beautiful homeland of East Prussia. If he had seen "my" oryx antelopes that day, perhaps he might have added a few more things which "fill the mind with increasing admiration" the more and the longer we think about them. But perhaps he might not have done so, either. There is a strange relationship between the "heroes of the human intellect" and animals. Albert Schweitzer once characterized it by comparing the attitude of many philosophers toward animals to the behavior of a cleaning lady with respect to the house dog. After a maid has carefully cleaned a room, she may close the door tightly and even take more precautions to keep the dog from entering it, since he might put tracks on the shiny floor or otherwise cause a disturbance. In a comparable way, most human thinkers, having established their philosophical systems, carefully keep the animals out of them, for very similar reasons.

APPENDIX

A. Common and Scientific Names of the Wild Mammals and Birds Mentioned in the Text

(*Note:* An asterisk indicates a species not occurring in the Serengeti National Park.)

Aardvark, *Orycteropus afer*
Aardwolf, *Proteles cristatus*
African buffalo, *Syncerus caffer*
African hare, *Lepus capensis*
African wild cat, *Felis libyca*
Baboon. *See* Olive baboon.
Banded mongoose, *Mungos mungo*
Bat-eared fox, *Otocyon megalotis*
Beisa antelope. *See* East African oryx antelope.
Black-backed jackal, *Canis mesomelas*
Black rhinoceros, *Diceros bicornis*
Blesbok, *Damaliscus dorcas phillipsi**
Blue wildebeest, *Connochaetes taurinus. See also* White-bearded wildebeest.
Boehm's zebra, *Equus quagga boehmi. See also* Plains zebra.
Bohor reedbuck, *Redunca redunca*
Brindled wildebeest. *See* Blue wildebeest.
Bush baby. *See* Lesser galago.
Bushbuck, *Tragelaphus scriptus*
Bush cat. *See* African wild cat.
Cape buffalo. *See* African buffalo.
Cape hare. *See* African hare.
Cat-related predators, family Felidae
Cheetah, *Acinonyx jubatus*
Coke's hartebeest. *See* Kongoni.
Common jackal, *Canis aureus*
Crowned guinea-fowl, *Numida meleagris*
Defassa waterbuck, *Kobus ellipsiprymnus defassa. See also* Waterbuck.
Dik-dik. *See* Kirk's dik-dik
Dorcas gazelle, *Gazella dorcas**
Duikers, subfamily Cephalophinae
East African oryx antelope, *Oryx beisa. See also* Fringe-eared oryx.
Eland antelope, *Tragelaphus (Taurotragus) oryx*
Fringe-eared oryx, *Oryx beisa callotis. See also* East African oryx antelope.
Gazelles, genus *Gazella*

Genets, *Genetta genetta* and *Genetta tigrina*
Giraffe, *Giraffa camelopardalis*
Gnu. *See* Wildebeest.
Golden jackal. *See* Common jackal.
Grant's gazelle, *Gazella (Nanger) granti. See also* Wide-horned Grant's gazelle.
Grant's zebra. *See* Boehm's zebra.
Greater kudu, *Tragelaphus strepsiceros**
Guinea-fowl. *See* Crowned guinea-fowl.
Hare. *See* African hare.
Hartebeest, *Alcelaphus buselaphus. See also* Kongoni.
Honey badger, *Mellivora capensis*
Hook-lipped rhinoceros. *See* Black rhinoceros.
Horned ungulates, family Bovidae
Hyena. *See* Spotted hyena.
Impala, *Aepyceros melampus*
Jackals. *See* Common jackal; Black-backed jackal.
Jumping hare, *Pedetes capensis*
Kirk's dik-dik, *Madoqua (Rhynchotragus) kirki*
Klipspringer, *Oreotragus oreotragus*
Kongoni, *Alcelaphus buselaphus cokei. See also* Hartebeest.
Kori bustard, *Otis kori*
Kudu. *See* Greater kudu.
Leopard, *Panthera pardus*
Lesser galago, *Galago senegalensis*
Lion, *Panthera leo*
Moose, *Alces alces**
Musk ox, *Ovibos moschatus**
Nilgai, Nilgai antelope, *Boselaphus tragocamelus**
Olive baboon, *Papio anubis*
Oryx, Oryx antelope. *See* East African oryx antelope.
Oryx antelopes, genus *Oryx*
Ostrich, *Struthio camelus*
Plains zebra, *Equus quagga. See also* Boehm's zebra.
Porcupine, *Hystrix cristata*
Reedbuck. *See* Bohor reedbuck.
Rhino, Rhinoceros. *See* Black rhinoceros.
Roan, Roan antelope, *Hippotragus equinus*
Robert's gazelle. *See* Wide-horned Grant's gazelle.
Sable antelope, *Hippotragus niger**
Serval, *Leptailurus serval*
Spotted hyena, *Crocuta crocuta*
Spring hare. *See* Jumping hare.
Steenbok, *Raphicerus campestris*
Thomson's gazelle, *Gazella thomsoni*
Topi, Topi antelope, *Damaliscus lunatus topi*
Tree dassie. *See* Tree hyrax.
Tree hyrax, *Dendrohyrax arboreus*
Warthog, *Phacochoerus aethiopicus*
Waterbuck, *Cobus ellipsiprymnus. See also* Defassa waterbuck.
White-bearded wildebeest, *Connochaetes taurinus albojubatus. See also* Blue wildebeest.
Wide-horned Grant's gazelle, *Gazella granti robertsi. See also* Grant's gazelle.
Wild cat. *See* African wild cat.

B. Author's Scientific Publications Resulting Largely or Exclusively from His Research in the Serengeti National Park, Tanzania

Walther, F. R. 1968. *Verhalten der Gazellen.* Wittenberg-Lutherstadt: A. Ziemsen Verlag. 144 pages.

______. 1969. Flight behaviour and avoidance of predators in Thomson's gazelle (*Gazella thomsoni* Günther 1884). *Behaviour* 34: 184–221.

______. 1972. Social grouping in Grant's gazelle (*Gazella granti* Brooke, 1872) in the Serengeti National Park. *Z. Tierpsychol.* 31: 348–403.

______. 1972. Territorial behaviour in certain horned ungulates with special reference to the examples of Thomson's and Grant's gazelles. *Zoologica Africana* 7: 303–307.

______. 1972. Horned ungulates. In: *Grzimek's Animal Life Encyclopedia,* vol. 13, pp. 272–307. New York: Van Nostrand Reinhold.

______. 1972. Duikers, dwarf antelopes and Tragelaphinae. In: *Grzimek's Animal Life Encyclopedia,* vol. 13, pp. 307–330. New York: Van Nostrand Reinhold.

______. 1972. Hartebeests, roan and sable antelopes, and waterbucks. In: *Grzimek's Animal Life Encyclopedia,* vol. 13, pp. 399–430. New York: Van Nostrand Reinhold.

______. 1972. The gazelles and their relatives. In: *Grzimek's Animal Life Encyclopedia,* vol. 13, pp. 431–449. New York: Van Nostrand Reinhold.

______. 1973. Round-the-clock activity of Thomson's gazelle (*Gazella thomsoni* Günther 1884) in the Serengeti National Park. *Z. Tierpsychol.* 32: 75–105.

______. 1973. On age class recognition and individual identification of Thomson's gazelle in the field. *J. South Afr. Wildlife Ass.* 2: 9–15.

______. 1974. Some reflections on expressive behaviour in combats and courtship of certain horned ungulates. In: *The Behaviour of Ungulates and Its Relation to Management,* ed. V. Geist and F. R. Walther, pp. 56–106. IUCN Publ. No. 24. Morges: IUCN.

______. 1975. Effects of intraspecific aggression on social organization, spatial distribution, and movements in certain East African horned ungulates. *Serengeti Research Inst., Annual Report 1974–75,* pp. 94–125.

______. 1977. Sex and activity dependency of distances between Thomson's gazelle (*Gazella thomsoni* Günther 1884). *Animal Behaviour* 25: 713–719.

______. 1977. Artiodactyla. In: *How Animals Communicate,* ed. T. A. Sebeok, pp. 655–714. Bloomington: Indiana University Press.

______. 1978. Aggressive behavior in captive and wild oryx antelope. *AAZPA Reg. Workshop Proc. 1977–78,* pp. 122–151.

______. 1978. Behavioral observations on oryx antelope (*Oryx beisa*) invading Serengeti National Park, Tanzania. *J. Mammal.* 59: 243–260.

______. 1978. Quantitative and functional variations of certain behaviour patterns in male Thomson's gazelle of different social status. *Behaviour* 65: 212–240.

______. 1978. Forms of aggression in Thomson's gazelle; their situational motivation and their relative frequency in different sex, age, and social classes. *Z. Tierpsychol.* 47: 113-172.

______. 1978. Mapping the structure and the marking system of a territory of the Thomson's gazelle. *East Afric. Wildlife J.* 16: 167-176.

______. 1979. Das Verhalten der Hornträger (Bovidae). *Handb. Zool.* 8, 10(30): 1-184.

______. 1984. *Communication and expression in hoofed mammals.* Bloomington: Indiana University Press. 423 pages.

______. 1991. On herding behavior. *Appl. Anim. Behav. Science* 29: 5-13.

______. 1992. Beobachtungen zur Situationseinsicht und zum Verhalten von Oryxantilopen in ziehenden Verbänden. *Zoolog. Garten* 62: 297-338.

______, E. Cary Mungall, and G. A. Grau. 1983. *Gazelles and their relatives—a study in territorial behavior.* Park Ridge, N.J.: Noyes Publ. 239 pages.

FRITZ R. WALTHER, now retired, was formerly a professor in the Department of Wildlife and Fisheries Sciences at Texas A & M University and has worked in several European zoos and African national parks. The author of *Communication and Expression in Hoofed Mammals* and numerous scientific papers, he has made many trips to Africa to study animal behavior, especially that of gazelles.